Mastering Python for Finance
Second Edition

Implement advanced state-of-the-art financial statistical applications using Python

James Ma Weiming

BIRMINGHAM - MUMBAI

Mastering Python for Finance
Second Edition

Commissioning Editor: Sunith Shetty
Acquisition Editor: Devika Battike
Content Development Editor: Snehal Kolte
Technical Editor: Manikandan Kurup
Copy Editor: Safis Editing
Project Coordinator: Manthan Patel
Proofreader: Safis Editing
Indexer: Tejal Daruwale Soni
Graphics: Jisha Chirayil
Production Coordinator: Arvindkumar Gupta

First published: April 2015
Second edition: April 2019

Production reference: 1300419

Published by Packt Publishing Ltd.
Livery Place
35 Livery Street
Birmingham
B3 2PB, UK.

ISBN 978-1-78934-646-6

www.packtpub.com

To the team at Skyline Trading on the CBOT trading floor - Mr. Milt Robinson, Brian Hickman, and Frank.

To my family, friends, and colleagues.

And, of course, to Paik Yen.

Contributors

About the author

James Ma Weiming is a software engineer based in Singapore. His studies and research are focused on financial technology, machine learning, data sciences, and computational finance. James started his career in financial services working with treasury fixed income and foreign exchange products, and fund distribution. His interests in derivatives led him to Chicago, where he worked with veteran traders of the Chicago Board of Trade to devise high-frequency, low-latency strategies to game the market. He holds an MS degree in finance from Illinois Tech's Stuart School of Business in the United States and a bachelor's degree in computer engineering from Nanyang Technological University.

About the reviewers

Anil Omanwar has over 11 years' experience in researching cognitive computing, while natural language processing (NLP), machine learning, information visualization, and text analytics are a few key areas of his research interests. He is proficient in sentiment analysis, questionnaire-based feedback, text clustering, and phrase extraction in diverse domains such as banking, oil and gas, life sciences, manufacturing, retail, social media, and others. He is currently associated with IBM Australia for NLP and IBM Watson in the oil and gas domain. The objective of his research is to automate critical manual decisions and assist domain experts to optimize human-machine capabilities. He holds multiple patents on emerging technologies, including NLP automation and device intelligence.

Rahul Shendge has a bachelor's degree in computer engineering from University of Pune and is certified in multiple technologies. He is an open source enthusiast and works as a senior software engineer. He has worked in multiple domains, including finance, healthcare, and education. He has hands-on experience of cloud, designing, and trading algorithms with machine learning. He is constantly exploring technical novelties and he is open-minded and eager to learn about new technologies. He is passionate about helping clients to make valuable business decisions using analytics in respective areas. His interests include working with and exploring data analytics solutions.

Packt is searching for authors like you

If you're interested in becoming an author for Packt, please visit `authors.packtpub.com` and apply today. We have worked with thousands of developers and tech professionals, just like you, to help them share their insight with the global tech community. You can make a general application, apply for a specific hot topic that we are recruiting an author for, or submit your own idea.

mapt.io

Mapt is an online digital library that gives you full access to over 5,000 books and videos, as well as industry leading tools to help you plan your personal development and advance your career. For more information, please visit our website.

Why subscribe?

- Spend less time learning and more time coding with practical eBooks and Videos from over 4,000 industry professionals

- Improve your learning with Skill Plans built especially for you

- Get a free eBook or video every month

- Mapt is fully searchable

- Copy and paste, print, and bookmark content

Packt.com

Did you know that Packt offers eBook versions of every book published, with PDF and ePub files available? You can upgrade to the eBook version at www.packt.com and as a print book customer, you are entitled to a discount on the eBook copy. Get in touch with us at customercare@packtpub.com for more details.

At www.packt.com, you can also read a collection of free technical articles, sign up for a range of free newsletters, and receive exclusive discounts and offers on Packt books and eBooks.

Table of Contents

Preface

This second edition of *Mastering Python for Finance* will guide you through carrying out complex financial calculations practiced in the finance industry, using next-generation methodologies. You will master the Python ecosystem by leveraging publicly available tools to successfully perform research studies and modeling, and learn how to manage risks using advanced examples.

You will start by setting up a Jupyter notebook to implement the tasks throughout the book. You will learn how to make efficient and powerful data-driven financial decisions using popular libraries such as TensorFlow, Keras, NumPy, SciPy, scikit-learn, and so on. You will also learn how to build financial applications by mastering concepts such as stocks, options, interest rates and their derivatives, and risk analytics using computational methods. With these foundations, you will learn how to apply statistical analysis on time series data and understand how to harness high-frequency data to devise trading strategies in building an algorithmic trading platform. You will learn to validate your trading strategies by implementing an event-driven backtesting system and measure its performance. Finally, you will explore machine learning and deep learning techniques that are applied in finance.

By the end of this book, you will have learned how to apply Python to different paradigms in the financial industry and perform efficient data analysis.

Who this book is for

If you are a financial or data analyst, or a software developer in the financial industry, who is interested in using advanced Python techniques for quantitative methods, this is the book you need! You will also find this book useful if you want to extend the functionalities of your existing financial applications using smart machine learning techniques.

What this book covers

Chapter 1, *Overview of Financial Analysis with Python*, goes briefly through setting up a Python environment, including a Jupyter Notebook, so that you can proceed with the rest of the chapters in this book. Within Jupyter, we will perform some time series analysis with pandas, using plots for analysis.

Chapter 2, *The Importance of Linearity in Finance,* uses Python to solve systems of linear equations, perform integer programming, and apply matrix algebra to the linear optimization of portfolio allocation.

Chapter 3, *Nonlinearity in Finance,* explores some methods that will help us extract information from nonlinear models. You will learn root-finding methods in nonlinear volatility modeling. The optimize module of SciPy contains the root and fsolve functions, which can also help us to perform root finding on non-linear models.

Chapter 4, *Numerical Methods for Pricing Options,* explores trees, lattices, and finite differencing schemes for the valuation of options.

Chapter 5, *Modeling Interest Rates and Derivatives,* discusses the bootstrapping process of the yield curve and covers some short-rate models for pricing interest rate derivatives with Python.

Chapter 6, *Statistical Analysis of Time Series Data,* introduces principal component analysis for identifying principal components. The Dicker-Fuller test is used for testing whether a time series is stationary.

Chapter 7, *Interactive Financial Analytics with VIX,* discusses volatility indexes. We will perform analytics on a US stock index and VIX data, and replicate the main index using the options prices of the sub-indexes.

Chapter 8, *Building an Algorithmic Trading Platform,* takes a step-by-step approach to developing a mean-reverting and trend-following live trading infrastructure using a broker API.

Chapter 9, *Implementing a Backtesting System,* discusses how to design and implement an event-driven backtesting system and helps you to visualize the performance of our simulated trading strategy.

Chapter 10, *Machine Learning for Finance,* introduces us to machine learning, allowing us to study its concepts and applications in finance. We will also look at some practical examples for applying machine learning to assist in trading decisions.

Chapter 11, *Deep Learning for Finance,* encourages us to take a hands-on approach to learning TensorFlow and Keras by building deep learning prediction models using neural networks.

To get the most out of this book

Prior experience in Python is required.

Download the example code files

You can download the example code files for this book from your account at www.packt.com. If you purchased this book elsewhere, you can visit www.packt.com/support and register to have the files emailed directly to you.

You can download the code files by following these steps:

1. Log in or register at www.packt.com.
2. Select the **SUPPORT** tab.
3. Click on **Code Downloads & Errata**.
4. Enter the name of the book in the **Search** box and follow the onscreen instructions.

Once the file is downloaded, please make sure that you unzip or extract the folder using the latest version of:

- WinRAR/7-Zip for Windows
- Zipeg/iZip/UnRarX for Mac
- 7-Zip/PeaZip for Linux

The code bundle for the book is also hosted on GitHub at https://github.com/PacktPublishing/Mastering-Python-for-Finance-Second-Edition. In case there's an update to the code, it will be updated on the existing GitHub repository.

We also have other code bundles from our rich catalog of books and videos available at https://github.com/PacktPublishing/. Check them out!

Download the color images

We also provide a PDF file that has color images of the screenshots/diagrams used in this book. You can download it here: http://www.packtpub.com/sites/default/files/downloads/9781789346466_ColorImages.pdf.

Conventions used

There are a number of text conventions used throughout this book.

CodeInText: Indicates code words in text, database table names, folder names, filenames, file extensions, pathnames, dummy URLs, user input, and Twitter handles. Here is an example: "By default, pandas' .plot() command uses the matplotlib library to display graphs."

A block of code is set as follows:

```
In [ ]:
    %matplotlib inline
    import quandl

    quandl.ApiConfig.api_key = QUANDL_API_KEY
    df = quandl.get('EURONEXT/ABN.4')
    daily_changes = df.pct_change(periods=1)
    daily_changes.plot();
```

When we wish to draw your attention to a particular part of a code block, the relevant lines or items are set in bold:

```
2015-02-26 TICK WIKI/AAPL open: 128.785 close: 130.415
2015-02-26 FILLED BUY 1 WIKI/AAPL at 128.785
2015-02-26 POSITION value:-128.785 upnl:1.630 rpnl:0.000
2015-02-27 TICK WIKI/AAPL open: 130.0 close: 128.46
```

Any command-line input or output is written as follows:

```
$ cd my_project_folder
$ virtualenv my_env
```

Bold: Indicates a new term, an important word, or words that you see onscreen. For example, words in menus or dialog boxes appear in the text like this. Here is an example: "To start your first notebook, select **New**, then **Python 3**."

 Warnings or important notes appear like this.

 Tips and tricks appear like this.

Get in touch

Feedback from our readers is always welcome.

General feedback: If you have questions about any aspect of this book, mention the book title in the subject of your message and email us at customercare@packtpub.com.

Errata: Although we have taken every care to ensure the accuracy of our content, mistakes do happen. If you have found a mistake in this book, we would be grateful if you would report this to us. Please visit www.packt.com/submit-errata, selecting your book, clicking on the Errata Submission Form link, and entering the details.

Piracy: If you come across any illegal copies of our works in any form on the Internet, we would be grateful if you would provide us with the location address or website name. Please contact us at copyright@packt.com with a link to the material.

If you are interested in becoming an author: If there is a topic that you have expertise in and you are interested in either writing or contributing to a book, please visit authors.packtpub.com.

Reviews

Please leave a review. Once you have read and used this book, why not leave a review on the site that you purchased it from? Potential readers can then see and use your unbiased opinion to make purchase decisions, we at Packt can understand what you think about our products, and our authors can see your feedback on their book. Thank you!

For more information about Packt, please visit packt.com.

Section 1: Getting Started with Python

This section will help us to set up Python on our machine in preparation for running code examples in this book.

This section will contain only one chapter:

- Chapter 1, *Overview of Financial Analysis with Python*

Overview of Financial Analysis with Python

1

Since the publication of my previous book *Mastering Python for Finance*, there have been significant upgrades to Python itself and many third-party libraries. Many tools and features have been deprecated in favor of new ones. This chapter walks you through how to get the latest tools available and how to prepare the environment that will be used throughout the rest of the book.

We will be using Quandl for the majority of datasets covered in this book. Quandl is a platform that serves financial, economic, and alternative data. These sources of data are contributed by various data publishers, including the United Nations, World Bank, central banks, trading exchanges, investment research firms, and even members of the Quandl community. With the Python Quandl module, you can easily download datasets and perform financial analytics to derive useful insights.

We will explore time series data manipulation using the `pandas` module. The two primary data structures in `pandas` are the Series object and the DataFrame object. Together, they can be used to plot charts and visualize complex information. Common methods of financial time series computation and analysis will be covered in this chapter.

The intention of this chapter is to serve as a foundation for setting up your working environment with libraries that will be used throughout this book. Over the years, like any software packages, the `pandas` module has evolved drastically with many breaking changes. Codes written years ago interfacing with older version of `pandas` will no longer work as many methods have been deprecated. The version of `pandas` used in this book is 0.23. Code written in this book conforms to this version of `pandas`.

In this chapter, we will cover the following:

- Setting up Python, Jupyter, Quandl, and other libraries for your environment
- Downloading datasets from Quandl and plotting your first chart
- Plotting last prices, volumes, and candlestick charts
- Calculating and plotting daily percentage and cumulative returns
- Plotting volatility, histograms, and Q-Q plots
- Visualizing correlations and generating the correlation matrix
- Visualizing simple moving averages and exponential moving averages

Getting Python

At the time of writing, the latest Python version is 3.7.0. You may download the latest version for Windows, macOS X, Linux/UNIX, and other operating systems from the official Python website at `https://www.python.org/downloads/`. Follow the installation instructions to install the base Python interpreter on your operating system.

The installation process should add Python to your environment path. To check the version of your installed Python, type the following command into the terminal if you are using macOS X/Linux, or the command prompt on Windows:

```
$ python --version
Python 3.7.0
```

For easy installation of Python libraries, consider using an all-in-one Python distribution such as Anaconda (`https://www.anaconda.com/download/`), Miniconda (`https://conda.io/miniconda.html`), or Enthought Canopy (`https://www.enthought.com/product/enthought-python-distribution/`). Advanced users, however, may prefer to control which libraries get installed with their base Python interpreter.

Preparing a virtual environment

At this point, it is advisable to set up a Python virtual environment. Virtual environments allow you to manage separate package installations that you need for a particular project, isolating the packages installed in other environments.

To install the virtual environment package in your terminal window, type the following:

```
$ pip install virtualenv
```

 On some systems, Python 3 may use a different `pip` executable and may need to be installed via an alternate `pip` command; for example: `$ pip3 install virtualenv`.

To create a virtual environment, go to your project's directory and run `virtualenv`. For example, if the name of your project folder is `my_project_folder`, type the following:

```
$ cd my_project_folder
$ virtualenv my_venv
```

`virtualenv my_venv` will create a folder in the current working directory that includes Python executable files of your base Python interpreter installed earlier, and a copy of the `pip` library, which you can use to install other packages.

Before using the new virtual environment, it needs to be activated. In a macOS X or Linux terminal, type the following command:

```
$ source my_venv/bin/activate
```

On Windows, the activation command is as follows:

```
$ my_project_folder\my_venv\Scripts\activate
```

The name of the current virtual environment will now appear on the left of the prompt (for example, `(my_venv) current_folder$`) to let you know that the selected Python environment is activated. Package installations from the same terminal window will be placed in the `my_venv` folder, isolated from the global Python interpreter.

 Virtual environments can help prevent conflicts should you have multiple applications using the same module but from different versions. This step (creating a virtual environment) is entirely optional as you can still use your default base interpreter to install packages.

Running Jupyter Notebook

Jupyter Notebook is a browser-based interactive computational environment for creating, executing, and visualizing interactive data across various programming languages. It was formerly known as **IPython** Notebook. IPython continues to exist as a Python shell and a kernel for Jupyter. Jupyter is an open-source software, free for all to use and learn about a variety of topics, from basic programming to advanced statistics or quantum mechanics.

To install Jupyter, type the following command in your terminal window:

```
$ pip install jupyter
```

Once installed, start Jupyter with the following command:

```
$ jupyter notebook
...
Copy/paste this URL into your browser when you connect for the first time,
to login with a token:
    http://127.0.0.1:8888/?token=27a16ee4d6042a53f6e31161449efcf7e71418f23e17549d
```

Watch your terminal window. When Jupyter has started, the console will provide information about this running status. You should also see a URL. Copy that URL into a web browser to bring you to the Jupyter computing interface.

Since Jupyter starts in the directory where you have issued the preceding command, Jupyter will list all saved notebooks in the working directory. If this is the first time you are working in the directory, the list will be empty.

To start your first notebook, select **New**, then **Python 3**. A new Jupyter Notebook will open in a new window. Henceforth, most computations in this book will be performed in Jupyter.

The Python Enhancement Proposal

Any design considerations in the Python programming language are documented as a **Python Enhancement Proposal** (**PEP**). Hundreds of PEPs have been written down, but probably the one that you should be familiar with is **PEP 8**, a style guide for Python developers to write better, readable code. The official repository for PEPs is `https://github.com/python/peps`.

What is a PEP?

PEPs are a numbered collection of design documents describing a feature, process, or environment related to Python. Each PEP is carefully maintained in a text file, containing technical specifications of a particular feature and its rationale for its existence. For example, PEP 0 serves as the index of all PEPs, while PEP 1 provides the purpose and guidelines of PEPs. As software developers, we often read code more than we write code. To create clear, concise, and readable code, we should always use a style guide as a coding convention. PEP 8 is a set of style guidelines for writing presentable Python code. You can read more about PEP 8 at `https://www.python.org/dev/peps/pep-0008/`.

The Zen of Python

PEP 20 embodies the Zen of Python, which is a collection of 20 software principles that guide the design of the Python programming language. To display this Easter egg, type the following command in your Python shell:

```
>> import this
The Zen of Python, by Tim Peters

Beautiful is better than ugly.
Explicit is better than implicit.
Simple is better than complex.
Complex is better than complicated.
Flat is better than nested.
Sparse is better than dense.
Readability counts.
Special cases aren't special enough to break the rules.
Although practicality beats purity.
Errors should never pass silently.
Unless explicitly silenced.
In the face of ambiguity, refuse the temptation to guess.
There should be one-- and preferably only one --obvious way to do it.
Although that way may not be obvious at first unless you're Dutch.
Now is better than never.
Although never is often better than *right* now.
If the implementation is hard to explain, it's a bad idea.
If the implementation is easy to explain, it may be a good idea.
Namespaces are one honking great idea -- let's do more of those!
```

Only 19 of the 20 aphorisms are shown. Can you figure out what is the last one? I leave it to your imagination!

Introduction to Quandl

Quandl is a platform that serves financial, economic, and alternative data. These sources of data are contributed by various data publishers, including the United Nations, World Bank, central banks, trading exchanges, and investment research firms.

With the Python Quandl module, you can easily get financial datasets into Python. Quandl offers free datasets, some of which are samples. Paid access is required for access to premium data products.

Setting up Quandl for your environment

The `Quandl` package requires the latest versions of NumPy and `pandas`. Additionally, we will require `matplotlib` for the rest of this chapter.

To install these packages, type the following code in your terminal window:

```
$ pip install quandl numpy pandas matplotlib
```

Over the years, there have been many changes to the `pandas` library. Code written for older versions of `pandas` may not work with the latest versions as there have been many deprecations. The version of `pandas` that we will be working with is 0.23. To check which version of `pandas` you are using, type the following command in a Python shell:

```
>>> import pandas
>>> pandas.__version__
'0.23.3'
```

An **API** (short for **Application Programming Interface**) key is required when using Quandl to request for datasets.

If you do not have a Quandl account, go through the following steps:

1. Open your browser and enter `https://www.quandl.com` in the address bar. This will display the following page:

2. Select **SIGN UP** and follow the instructions to create a free account. Your API key will be shown after you have successfully registered.
3. Copy this key and keep it safe elsewhere as you will need this it later. Otherwise, you may retrieve this key again in your **ACCOUNT SETTINGS**.
4. Remember to check your email inbox for a welcome message and verify your Quandl account, as continued use of the API key requires a verified and valid Quandl account.

 Anonymous users have a limit of 20 calls per 10 minutes and 50 calls per day. Authenticated free users have a limit of 300 calls per 10 seconds, 2,000 calls per 10 minutes, and a limit of 50,000 calls per day.

Plotting a time series chart

A simple and effective technique for analyzing time series data is by visualizing it on a graph, from which we can infer certain assumptions. This section will guide you through the process of downloading a dataset of stock prices from Quandl and plotting it on a price and volume graph. We will also cover plotting candlestick charts, which will give us more information than line charts.

Retrieving datasets from Quandl

Fetching data from Quandl into Python is fairly straightforward. Suppose we are interested in ABN Amro Group from the Euronext Stock Exchange. The ticker symbol in Quandl is EURONEXT/ABN. In a Jupyter notebook cell, run the following command:

```
In [ ]:
    import quandl

    # Replace with your own Quandl API key
    QUANDL_API_KEY = 'BCzkk3NDWt7H9yjzx-DY'
    quandl.ApiConfig.api_key = QUANDL_API_KEY
    df = quandl.get('EURONEXT/ABN')
```

 It is a good practice to store your Quandl API key in a constant variable. This way, should your API key change, you only need to update it in one place!

After importing the quandl package, we store our Quandl API key in the constant variable, QUANDL_API_KEY, which will be reused in the rest of this chapter. This constant value is used to set the Quandl module API key, and only needs to be executed once for every import of the quandl package. The quandl.get() method on the next line is called to download the ABN dataset from Quandl right into our df variable. Note that EURONEXT is an abbreviation for the data provider, Euronext Stock Exchange.

By default, Quandl will retrieve the dataset into a `pandas` DataFrame. We can inspect the head and tail of the DataFrame as follows:

```
In [ ]:
    df.head()
Out[ ]:
                Open   High    Low   Last     Volume     Turnover
    Date
    2015-11-20  18.18  18.43  18.000  18.35  38392898.0  7.003281e+08
    2015-11-23  18.45  18.70  18.215  18.61   3352514.0  6.186446e+07
    2015-11-24  18.70  18.80  18.370  18.80   4871901.0  8.994087e+07
    2015-11-25  18.85  19.50  18.770  19.45   4802607.0  9.153862e+07
    2015-11-26  19.48  19.67  19.410  19.43   1648481.0  3.220713e+07

In [ ]:
    df.tail()
Out[ ]:
                Open   High    Low   Last     Volume     Turnover
    Date
    2018-08-06  23.50  23.59  23.29  23.34   1126371.0  2.634333e+07
    2018-08-07  23.59  23.60  23.31  23.33   1785613.0  4.177652e+07
    2018-08-08  24.00  24.39  23.83  24.14   4165320.0  1.007085e+08
    2018-08-09  24.40  24.46  24.16  24.37   2422470.0  5.895752e+07
    2018-08-10  23.70  23.94  23.28  23.51   3951850.0  9.336493e+07
```

By default, the `head()` and `tail()` commands will display the first and last five rows of the DataFrame, respectively. You can define the number of rows to display by passing a number in its argument. For example, `head(100)` will show the first 100 rows in the DataFrame.

Without any additional parameters set for the `get()` method, the entire time series dataset is retrieved, dating from the previous business day all the way back to November 2015 on a daily basis.

To visualize this DataFrame, we can plot a graph using the `plot()` command:

```
In [ ]:
    %matplotlib inline
    import matplotlib.pyplot as plt

    df.plot();
```

The last command outputs a simple plot:

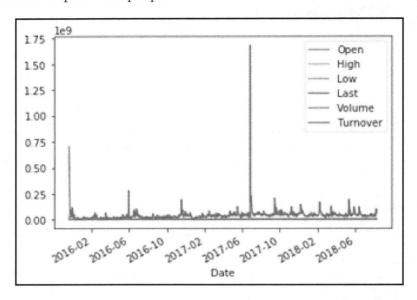

The `plot()` method of `pandas` returns an Axes object. A string representation of this object is printed on the console along with the `plot()` command. To suppress this information, we can add a semicolon (;) at the end of the last statement. Alternatively, we can add a `pass` statement at the bottom of the cell. Alternatively, assigning the plotting function to a variable also suppresses the output.

By default, the `plot()` command in `pandas` uses the `matplotlib` library to display graphs. If you are having errors, check to ensure this library is installed and `%matplotlib inline` is called once.

You can customize the look and feel of your charts. Further information on the `plot` command in the `pandas` DataFrame is available in the `pandas` documentation at `https://pandas.pydata.org/pandas-docs/stable/generated/pandas.DataFrame.plot.html`.

Plotting a price and volume chart

When no parameters are supplied to the `plot()` command, a line graph is plotted using all columns of the target DataFrame, on the same graph. This produces a cluttered view which does not give us much information. To effectively extract insights from this data, we can plot a financial graph of a stock with daily closing price relative to its trading volume. To facilitate this, type the following command:

```
In [ ]:
    prices = df['Last']
    volumes = df['Volume']
```

The preceding command stores our data of interest into the `closing_prices` and `volumes` variables, respectively. We can peek at the top and bottom rows of the resulting `pandas` Series data type with the `head()` and `tail()` commands:

```
In [ ]:
    prices.head()
Out[ ]:
    Date
    2015-11-20    18.35
    2015-11-23    18.61
    2015-11-24    18.80
    2015-11-25    19.45
    2015-11-26    19.43
    Name: Last, dtype: float64

In [ ]:
    volumes.tail()
Out[ ]:
    Date
    2018-08-03    1252024.0
    2018-08-06    1126371.0
    2018-08-07    1785613.0
    2018-08-08    4165320.0
    2018-08-09    2422470.0
    Name: Volume, dtype: float64
```

To find out the type of a particular variable, use the `type()` command. For example, `type(volumes)` produces `pandas.core.series.Series`, which tells us that the `volumes` variable is actually a `pandas` Series data type object.

Observe that data is available from 2018 all the way back to 2015. We can now plot the price and volume chart:

```
In [ ]:
    # The top plot consisting of daily closing prices
    top = plt.subplot2grid((4, 4), (0, 0), rowspan=3, colspan=4)
    top.plot(prices.index, prices, label='Last')
    plt.title('ABN Last Price from 2015 - 2018')
    plt.legend(loc=2)

    # The bottom plot consisting of daily trading volume
    bottom = plt.subplot2grid((4, 4), (3,0), rowspan=1, colspan=4)
    bottom.bar(volumes.index, volumes)
    plt.title('ABN Daily Trading Volume')

    plt.gcf().set_size_inches(12, 8)
    plt.subplots_adjust(hspace=0.75)
```

This produces the following graph:

On the first line, the `subplot2grid` command with the first parameter, `(4,4)`, divides the entire graph into a 4 x 4 grid. The second parameter `(0,0)` specifies that the given plot will be anchored on the top-left corner of the graph. The keyword parameter, `rowspan=3`, indicates the plot will occupy 3 of the 4 available rows on the grid, effectively as tall as 75% of the graph. The keyword parameter, `colspan=4`, indicates that the plot will occupy all 4 columns of the grid, using up all of its available width. The command returns a `matplotlib` axis object, which we will use to plot the upper portion of the graph.

On the second line, the `plot()` command renders the upper chart, with date and time values on the *x* axis, and prices on the *y* axis. In the next two lines, we specify the title of the current plot, along with a legend for the time series data placed in the upper-left corner.

Next, we perform the same actions to render the daily trading volume on the bottom chart, specifying a 1-row-by-4-column grid space anchored on the bottom-left corner of the graph.

In the `legend()` command, the `loc` keyword accepts an integer value as the location code of the legend. A value of `2` translates to an upper-left location. For a table of location codes, see the Legend documentation of `matplotlib` at `https://matplotlib.org/api/legend_api.html?highlight=legend#module-matplotlib.legend`.

To make our figure appear bigger, we invoke the `set_size_inches()` command to set the figure to 9 inches wide by 6 inches high, resulting in a rectangular-shaped figure. The preceding `gcf()` command simply means **get current figure**. Finally, the `subplots_adjust()` command with a `hspace` parameter is called to add a small amount of height between the top and bottom subplots.

The command `subplots_adjust()` tunes the subplot layout. Acceptable parameters are `left`, `right`, `bottom`, `top`, `wspace`, and `hspace`. For further information on these, see the `matplotlib` documentation at `https://matplotlib.org/api/_as_gen/matplotlib.pyplot.subplots_adjust.html`.

Plotting a candlestick chart

A candlestick chart is another type of popular financial chart that shows more information than just a single price. A candlestick represents a tick at each particular point of time with four important pieces of information: the open, the high, the low, and the close.

The `matplotlib.finance` module has been deprecated. Instead, we can use another package, `mpl_finance`, that consists of extracted code. To install this package, in your terminal window, type the following command:

```
$ pip install mpl-finance
```

To visualize the candles more closely, we will use a subset of the ABN dataset. In the following example, we query from Quandl the daily prices for the month of July 2018 as our dataset, and plot a candlestick chart, as follows:

```
In [ ]:
    %matplotlib inline
    import quandl
    from mpl_finance import candlestick_ohlc
    import matplotlib.dates as mdates
    import matplotlib.pyplot as plt

    quandl.ApiConfig.api_key = QUANDL_API_KEY
    df_subset = quandl.get('EURONEXT/ABN',
                            start_date='2018-07-01',
                            end_date='2018-07-31')

    df_subset['Date'] = df_subset.index.map(mdates.date2num)
    df_ohlc = df_subset[['Date','Open', 'High', 'Low', 'Last']]

    figure, ax = plt.subplots(figsize = (8,4))
    formatter = mdates.DateFormatter('%Y-%m-%d')
    ax.xaxis.set_major_formatter(formatter)
    candlestick_ohlc(ax,
                     df_ohlc.values,
                     width=0.8,
                     colorup='green',
                     colordown='red')
    plt.show()
```

This produces a candlestick chart as shown in the following screenshot:

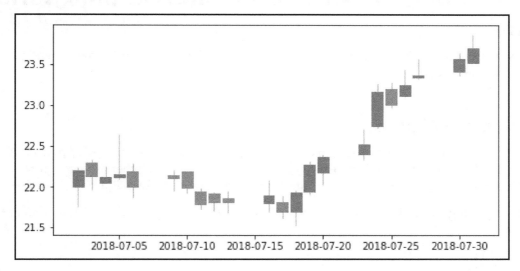

 You can specify the `start_date` and `end_date` parameters in the `quandl.get()` command to retrieve the dataset for the selected date range.

Prices retrieved from Quandl are placed in a variable named `df_dataset`. As the plot function of `matplotlib` requires its own formatting, the `mdates.date2num` command converts the index values containing the date and time, and places them in a new column named `Date`.

The candlestick's date, open, high, low, and close data columns are explicitly extracted as a DataFrame in the `df_ohlc` variable. `plt.subplots()` creates a plot figure with 8 inches wide and 4 inches high. Labels along the *x* axis are formatted into a human-readable format.

Our data is now ready for plotting in as a candlestick chart by calling the `candlestick_ohlc()` command, with a candlestick width of 0.8 (or 80% of a full day's width). Up ticks whose close price is higher than the open price are represented in green, while down ticks, whose close price are lower than the open price, are represented in red. Finally, we add the `plt.show()` command to display the candlestick chart.

Performing financial analytics on time series data

In this section, we will visualize some statistical properties of time series data used in financial analytics.

Plotting returns

One of the classic measures of security performance is its returns over a prior period. A simple method for calculating returns in `pandas` is `pct_change`, where the percentage change from the previous row is computed for every row in the DataFrame.

In the following example, we use ABN stock data to plot a simple graph of daily percentage returns:

```
In [ ]:
    %matplotlib inline
    import quandl

    quandl.ApiConfig.api_key = QUANDL_API_KEY
    df = quandl.get('EURONEXT/ABN.4')
    daily_changes = df.pct_change(periods=1)
    daily_changes.plot();
```

A line plot of daily percentage returns is shown as follows:

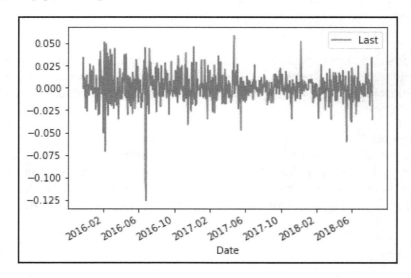

In the `quandl.get()` method, we postfix the ticker symbol with `.4` to specify the retrieval of only the fourth column of the dataset, which contains the last prices. In the call to `pct_change`, the `period` argument specifies the number of periods to shift to form the percentage change, which by default is `1`.

> Instead of using the postfix notation in the ticker symbol to specify the column of the dataset to download, we can pass the `column_index` parameter together with the index of the column. For example, `quandl.get('EURONEXT/ABN.4')` is the same as calling `quandl.get('EURONEXT/ABN', column_index=4)`.

Plotting cumulative returns

To find out how our portfolio has performed, we can sum its returns over a period of time. The `cumsum` method of `pandas` returns the cumulative sum over a DataFrame.

In the following example, we plot the cumulative sum of `daily_changes` of the ABN calculated previously:

```
In [ ]:
    df_cumsum = daily_changes.cumsum()
    df_cumsum.plot();
```

This gives us the following output graph:

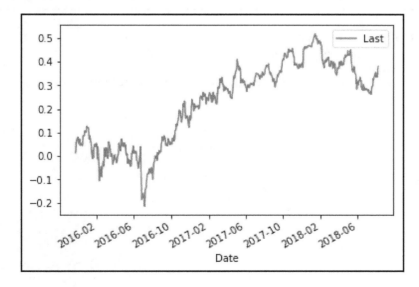

Plotting a histogram

Histograms tell us how distributed data is. In this example, we are interested in how distributed the daily returns of ABN are. We use the `hist()` method on a DataFrame with a bin size of 50:

```
In [ ]:
    daily_changes.hist(bins=50, figsize=(8, 4));
```

The histogram output is shown as follows:

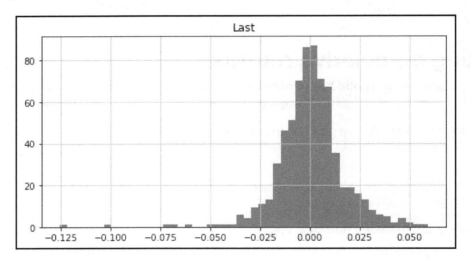

When there are multiple data columns in a `pandas` DataFrame, the `hist()` method will automatically plot each histogram on its own separate plot.

We can use the `describe()` method to summarize the central tendency, dispersion, and shape of a dataset's distribution:

```
In [ ]:
    daily_changes.describe()
Out[ ]:
                 Last
    count  692.000000
    mean     0.000499
    std      0.016701
    min     -0.125527
    25%     -0.007992
    50%      0.000584
    75%      0.008777
    max      0.059123
```

From the histogram, the returns tend to be distributed about the mean of 0.0, or 0.000499 to be exact. Besides this miniscule skew to the right, the data appears fairly symmetrical and normally distributed. The standard deviation is 0.016701. The percentiles tell us that 25% of the points fall below −0.007992, 50% below 0.000584, and 75% below 0.008777.

Plotting volatility

One way of analyzing the distribution of returns is measuring its standard deviation. **Standard deviation** is a measure of dispersion around the mean. A high standard deviation value for past returns indicates a high historical volatility of stock price movement.

The rolling() method of pandas helps us to visualize specific time series operations over a period of time. To calculate standard deviations of the percentage change of returns in our computed ABN dataset, we use the std() method, which returns a DataFrame or Series object that can be used to plot a chart. The following example illustrates this:

```
In [ ]:
    df_filled = df.asfreq('D', method='ffill')
    df_returns = df_filled.pct_change()
    df_std = df_returns.rolling(window=30, min_periods=30).std()
    df_std.plot();
```

This gives us the following volatility plot:

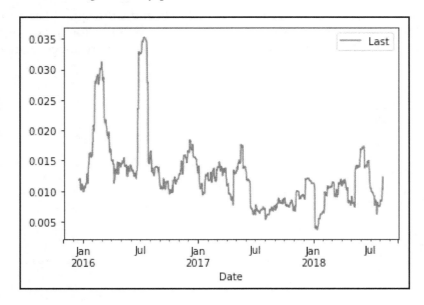

Our original time series datasets exclude weekends and public holidays, which must be taken into account when using the `rolling()` method. The `df.asfreq()` command will re-index time series data on a daily frequency, creating new indexes in place of missing ones. The `method` parameter with a value of `ffill` specifies that we will propagate the last valid observation forward in place of missing values during re-indexing.

In the `rolling()` command, we specified the `window` parameter with a value of 30, which is the number of observations used for calculating the statistic. In other words, the standard deviation of each period is calculated with a sample size of 30. Since the first 30 rows do not have a sample size that is enough to calculate the standard deviation, we can exclude these rows by specifying `min_periods` as 30.

The chosen value of 30 approximates the monthly standard deviation of returns. Note that choosing wider window periods represents less of the data being measured.

A quantile-quantile plot

A Q-Q (quantile-quantile) plot is a probability distribution plot, where the quantiles of two distributions are plotted against each other. If the distributions are linearly related, the points in the Q-Q plot will lie along a line. Compared to histograms, Q-Q plots help us to visualize points that lie outside the line for positive and negative skews, as well as excess kurtosis.

The `probplot()` of `scipy.stats` helps us to calculate and show quantiles for a probability plot. A best-fit line for the data is also drawn. In the following example, we use the last prices of the ABN stock dataset and compute the daily percentage change for charting a Q-Q plot:

```
In [ ]:
    %matplotlib inline
    import quandl
    from scipy import stats
    from scipy.stats import probplot

    quandl.ApiConfig.api_key = QUANDL_API_KEY
    df = quandl.get('EURONEXT/ABN.4')
    daily_changes = df.pct_change(periods=1).dropna()

    figure = plt.figure(figsize=(8,4))
    ax = figure.add_subplot(111)
    stats.probplot(daily_changes['Last'], dist='norm', plot=ax)
    plt.show();
```

This gives us the following Q-Q plot:

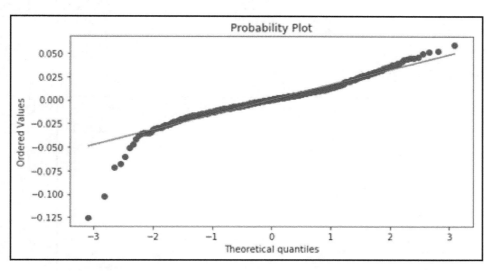

When all points fall exactly along the red line, the distribution of data implies perfect correspondences to a normal distribution. Most of our data is close to being perfectly correlated between quantiles -2 and +2. Outside this range, there begin to be differences in correlation of the distribution, with more negative skews at the tails.

Downloading multiple time series data

We pass a single Quandl code as a string object in the first parameter of the `quandl.get()` command to download a single dataset. To download multiple datasets, we can pass a list of Quandl codes.

In the following example, we are interested in the prices of three banking stocks—ABN Amro, Banco Santander, and Kas Bank. The two-year prices from 2016 to 2017 are stored in the `df` variable, with only the last prices downloaded:

```
In [ ]:
    %matplotlib inline
    import quandl

    quandl.ApiConfig.api_key = QUANDL_API_KEY
    df = quandl.get(['EURONEXT/ABN.4',
                     'EURONEXT/SANTA.4',
```

```
                           'EURONEXT/KA.4'],
                   collapse='monthly',
                   start_date='2016-01-01',
                   end_date='2017-12-31')
        df.plot();
```

The following plot is generated:

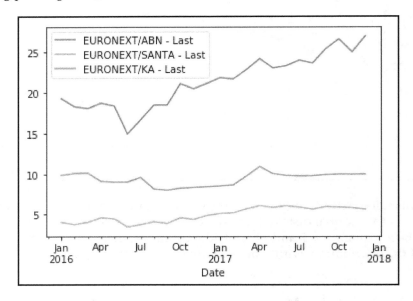

By default, `quandl.get()` returns daily prices. We may also specify other types of frequency for the dataset to download. In this example, we specified `collapse='monthly'` to download monthly prices.

Displaying the correlation matrix

Correlation is a statistical association of how closely two variables have a linear relationship with each other. We can perform a correlation calculation on the returns of two time series datasets to give us a value between -1 and 1. A correlation value of 0 indicates that the returns of the two time series have no relation to each other. A high correlation value close to 1 indicates that the returns of the two time series data tend to move together. A low value close to -1 indicates that returns tend to move inversely in relation to each other.

In `pandas`, the `corr()` method computes the correlations between columns in its supplied DataFrame and outputs these values as a matrix. In the previous example, we have three datasets available in the DataFrame `df`. To output the correlation matrix of returns, run the following command:

```
In [ ]:
    df.pct_change().corr()
Out[ ]:
                       EURONEXT/ABN - Last ... EURONEXT/KA - Last
    EURONEXT/ABN - Last           1.000000 ...           0.096238
    EURONEXT/SANTA - Last         0.809824 ...           0.058095
    EURONEXT/KA - Last            0.096238 ...           1.000000
```

From the correlation matrix output, we can infer that the ABN Amro and Banco Santander stocks are highly correlated during the two years from 2016 to 2017 with a value of `0.809824`.

By default, the `corr()` command uses the Pearson correlation coefficient to compute pairwise correlations. This is equivalent to calling `corr(method='pearson')`. Other valid values are `kendall` and `spearman` for the Kendall Tau and Spearman rank correlation coefficients, respectively.

Plotting correlations

Visualizing correlations can also be achieved with the `rolling()` command. We will use the Last prices of ABN and SANTA on a daily basis from 2016 to 2017, from Quandl. The two datasets are downloaded to the DataFrame `df`, and its rolling correlations plotted as follows:

```
In [ ]:
    %matplotlib inline
    import quandl

    quandl.ApiConfig.api_key = QUANDL_API_KEY
    df = quandl.get(['EURONEXT/ABN.4', 'EURONEXT/SANTA.4'],
                    start_date='2016-01-01',
                    end_date='2017-12-31')

    df_filled = df.asfreq('D', method='ffill')
    daily_changes= df_filled.pct_change()
    abn_returns = daily_changes['EURONEXT/ABN - Last']
    santa_returns = daily_changes['EURONEXT/SANTA - Last']
    window = int(len(df_filled.index)/2)
    df_corrs = abn_returns\
```

```
        .rolling(window=window, min_periods=window)\
        .corr(other=santa_returns)
        .dropna()
df_corrs.plot(figsize=(12,8));
```

The correlation plot is shown in the following screenshot:

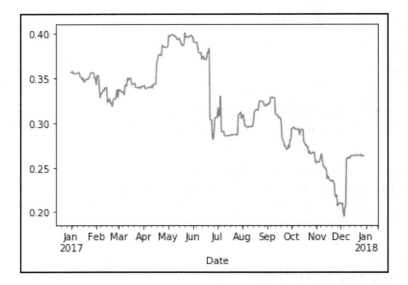

The `df_filled` variable contains a DataFrame with its index re-indexed on a daily frequency basis and missing values forward-filled in preparation for the `rolling()` command. The DataFrame, `daily_changes`, stores the daily percentage returns, and its columns are extracted into a separate Series object as `abn_returns` and `santa_returns`, respectively. The `window` variable stores the average number of days per year in the two-year dataset. This variable is supplied into the parameters of the `rolling()` command. The parameter `window` indicates we will perform a one-year rolling correlation. The `min_periods` parameter indicates that correlation will be calculated when only the full sample size is present for calculation. In this case, there are no correlation values for the first year in the `df_corrs` dataset. Finally, the `plot()` command displays the chart of one-year rolling correlations of daily returns throughout the year of 2017.

Simple moving averages

A common technical indicator for time series data analysis is moving averages. The `mean()` method can be used to compute the mean of values for a given window in the `rolling()` command. For example, a 5-day **Simple Moving Average** (**SMA**) is the average of prices for the last five trading days, computed daily over a time period. Similarly, we can also compute a longer term 30-day simple moving average. These two moving averages can be used together to generate crossover signals.

In the following example, we download the daily closing prices of ABN, compute the short- and long-term SMAs, and visualize them on a single plot:

```
In [ ]:
    %matplotlib inline
    import quandl
    import pandas as pd

    quandl.ApiConfig.api_key = QUANDL_API_KEY
    df = quandl.get('EURONEXT/ABN.4')

    df_filled = df.asfreq('D', method='ffill')
    df_last = df['Last']

    series_short = df_last.rolling(window=5, min_periods=5).mean()
    series_long = df_last.rolling(window=30, min_periods=30).mean()

    df_sma = pd.DataFrame(columns=['short', 'long'])
    df_sma['short'] = series_short
    df_sma['long'] = series_long
    df_sma.plot(figsize=(12, 8));
```

This produces the following plots:

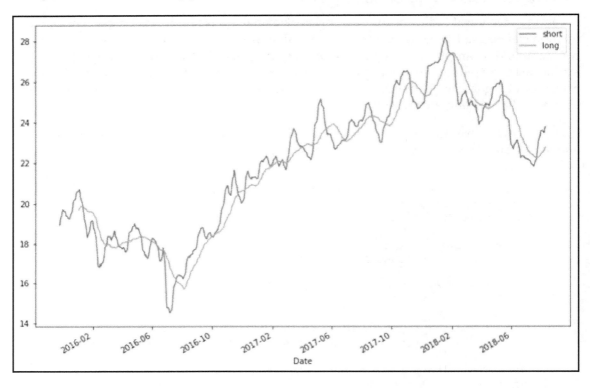

We use a 5-day average for the short-term SMA and 30 days for a long-term SMA. The `min_periods` parameter is supplied to exclude the first rows that do not have sufficient sample size for computing the SMA. The `df_sma` variable is a newly-created `pandas` DataFrame for storing SMA computations. We then plot a 12-inch-by-8-inch graph. From the graph, we can see a number of points where the short-term SMA intercepts the long-term SMA. Chartists use crossovers to identify trends and generate signals. The window periods of 5 and 10 are purely suggested values; you might tweak these values to find a suitable interpretation of your own.

Exponential moving averages

Another approach in the calculation of moving averages is the **Exponential Moving Average (EMA)**. Recall that the simple moving average assigns equal weight to prices within a window period. However, in EMA, the most recent prices are assigned a higher weight than older prices. This weight is assigned on an exponential basis.

The `ewm()` method of the `pandas` DataFrame provides exponential weighted functions. The `span` parameter specifies the window period for the decay behavior. The same ABN dataset with EMA is plotted as follows:

```
In [ ]:
    %matplotlib inline
    import quandl
    import pandas as pd

    quandl.ApiConfig.api_key = QUANDL_API_KEY
    df = quandl.get('EURONEXT/ABN.4')

    df_filled = df.asfreq('D', method='ffill')
    df_last = df['Last']

    series_short = df_last.ewm(span=5).mean()
    series_long = df_last.ewm(span=30).mean()

    df_sma = pd.DataFrame(columns=['short', 'long'])
    df_sma['short'] = series_short
    df_sma['long'] = series_long
    df_sma.plot(figsize=(12, 8));
```

This produces the following plot:

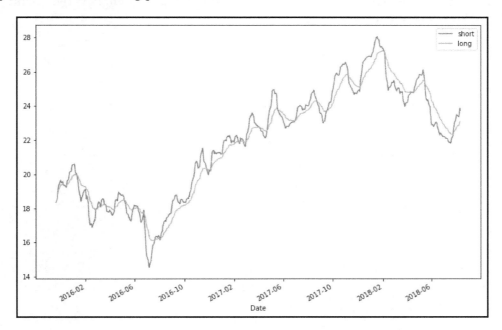

The chart patterns for the SMA and EMA are largely the same. Since EMAs place a higher weighting on recent data than on older data, they are more reactive to price changes than SMAs are.

 Besides varying window periods, you can experiment with combinations of SMA and EMA prices to derive more insights!

Summary

In this chapter, we set up our working environment with Python 3.7 and used the virtual environment package to manage separate package installations. The `pip` command is a handy Python package manager that easily downloads and installs Python modules, including Jupyter, Quandl, and `pandas`. Jupyter is a browser-based interactive computational environment for executing Python code and visualizing data. With a Quandl account, we can easily obtain high-quality time series datasets. These sources of data are contributed by various data publishers. Datasets directly download into a `pandas` DataFrame object that allows us to perform financial analytics, such as plotting daily percentage returns, histograms, Q-Q plots, correlations, simple moving averages, and exponential moving averages.

Section 2: Financial Concepts 2

This section covers financial concepts and mathematical models discussed by finance engineering practitioners.

In this section, we shall go through the following chapters:

The Importance of Linearity in Finance

2

Nonlinear dynamics play a vital role in our world. Linear models are often employed in economics due to being easier to study and their easier modeling capabilities. In finance, linear models are widely used to help price securities and perform optimal portfolio allocation, among other useful things. One significant aspect of linearity in financial modeling is its assurance that a problem terminates at a globally-optimal solution.

In order to perform prediction and forecasting, regression analysis is widely used in the field of statistics to estimate relationships among variables. With an extensive mathematics library being one of Python's greatest strength, Python is frequently used as a scientific scripting language to aid in these problems. Modules such as the SciPy and NumPy packages contain a variety of linear regression functions for data scientists to work with.

In traditional portfolio management, the allocation of assets follows a linear pattern, and investors have individual styles of investing. We can describe the problem of portfolio allocation as a system of linear equations, containing either equalities or inequalities. These linear systems can then be represented in a matrix form as $Ax=B$, where A is our known coefficient value, B is the observed result, and x is the vector of values that we want to find out. More often than not, x contains the optimal security weights to maximize our utility. Using matrix algebra, we can efficiently solve for x using either direct or indirect methods.

In this chapter, we will cover the following topics:

- Examining the Capital Asset Pricing Model and the security market line
- Solving for the security market line using regression
- Examining the APT model and performing a multivariate linear regression
- Understanding linear optimization in portfolio allocation
- Performing linear optimization using the Pulp package
- Understanding the outcomes of linear programming
- Introduction to integer programming
- Implementing a linear integer programming model with binary conditions
- Solving systems of linear equations with equalities using matrix linear algebra
- Solving systems of linear equations directly with LU, Cholesky, and QR decomposition
- Solving systems of linear equations indirectly with the Jacobi and Gauss-Seidel method

The Capital Asset Pricing Model and the security market line

A lot of the financial literature devotes exclusive discussions to the **Capital Asset Pricing Model** (**CAPM**). In this section, we will explore key concepts that highlight the importance of linearity in finance.

In the famous CAPM, the relationship between risk and rates of return in a security is described as follows:

$$R_i = R_f + \beta_i(E[R_{mkt}] - R_f)$$

For a security, *i*, its returns are defined as R_i and its beta as β_i. The CAPM defines the return of the security as the sum of the risk-free rate, R_f, and the multiplication of its beta with the risk premium. The risk premium can be thought of as the market portfolio's excess returns exclusive of the risk-free rate. The following is a visual representation of the CAPM:

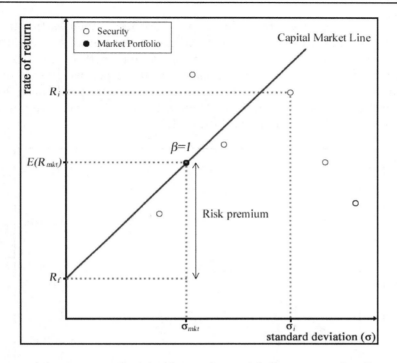

Beta is a measure of the systematic risk of a stock – a risk that cannot be diversified away. In essence, it describes the sensitivity of stock returns with respect to movements in the market. For example, a stock with a beta of zero produces no excess returns regardless of the direction the market moves in. It can only grow at a risk-free rate. A stock with a beta of 1 indicates that the stock moves perfectly with the market.

The beta is mathematically derived by dividing the covariance of returns between the stock and the market by the variance of the market returns.

The CAPM model measures the relationship between risk and stock returns for every stock in the portfolio basket. By outlining the sum of this relationship, we obtain combinations or weights of risky securities that produce the lowest portfolio risk for every level of portfolio return. An investor who wishes to receive a particular return would own one such combination of an optimal portfolio that provides the least risk possible. Combinations of optimal portfolios lie along a line called the **efficient frontier**.

Along the efficient frontier, there exists a tangent point that denotes the best optimal portfolio available and gives the highest rate of return in exchange for the lowest risk possible. This optimal portfolio at the tangent point is known as the **market portfolio**.

There exists a straight line drawn from the market portfolio to the risk-free rate. This line is called the **Capital Market Line** (**CML**). The CML can be thought of as the highest Sharpe ratio available among all the other Sharpe ratios of optimal portfolios. The **Sharpe ratio** is a risk-adjusted performance measure defined as the portfolio's excess returns over the risk-free rate per unit of its risk in standard deviations. Investors are particularly interested in holding combinations of assets along the CML line. The following diagram illustrates the efficient frontier, the market portfolio, and the CML:

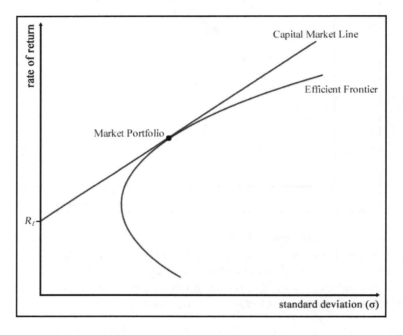

Another line of interest in CAPM studies is the **Security Market Line** (**SML**). The SML plots the asset's expected returns against its beta. For a security with a beta value of 1, its returns perfectly match the market's returns. Any security priced above the SML is deemed to be undervalued since investors expect a higher return given the same amount of risk. Conversely, any security priced below the SML is deemed to be overvalued, as follows:

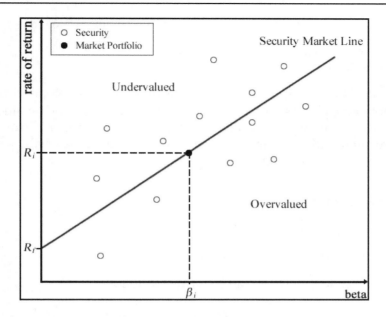

Suppose we are interested in finding the beta, β_i, of a security. We can regress the company's stock returns, R_i, against the market's returns, R_M, along with an intercept, α, in the form of the $R_i=\alpha+\beta R_M$ equation.

Consider the following set of stock return and market return data measured over five time periods:

Time period	Stock returns	Market returns
1	0.065	0.055
2	0.0265	-0.09
3	-0.0593	-0.041
4	-0.001	0.045
5	0.0346	0.022

Using the `stats` module of SciPy, we will perform a least-squares regression on the CAPM model, and derive the values of α and β_i by running the following code in Python:

```
In [ ]:
    """
    Linear regression with SciPy
    """
    from scipy import stats
```

```
stock_returns = [0.065, 0.0265, -0.0593, -0.001, 0.0346]
mkt_returns = [0.055, -0.09, -0.041, 0.045, 0.022]
beta, alpha, r_value, p_value, std_err = \
    stats.linregress(stock_returns, mkt_returns)
```

The `scipty.stats.linregress` function returns five values: the slope of the regression line, the intercept of the regression line, the correlation coefficient, the p-value for a hypothesis test with a null hypothesis of a zero slope, and the standard error of the estimate. We are interested in finding the slope and intercept of the line by printing the values of `beta` and `alpha`, respectively:

```
In [ ]:
    print(beta, alpha)
Out[ ]:
    0.5077431878770808 -0.008481900352462384
```

The beta of the stock is 0.5077 and the alpha is nearly zero.

The equation that describes the SML can be written as follows:

$$E[R_i] = R_f + \beta_i(E[R_M] - R_f)$$

The term $E[R_M] - R_f$ is the market risk premium, and $E[R_M]$ is the expected return on the market portfolio. R_f is the return on the risk-free rate, $E[R_i]$ is the expected return on asset, i, and β_i is the beta of the asset.

Suppose the risk-free rate is 5% and the market risk premium is 8.5%. What is the expected return of the stock? Based on the CAPM, an equity with a beta of 0.5077 would have a risk premium of 0.5077×8.5%, or 4.3%. The risk-free rate is 5%, so the expected return on the equity is 9.3%.

If the security is observed in the same time period to have a higher return (for example, 10.5%) than the expected stock return, the security can be said to be undervalued, since the investor can expect a greater return for the same amount of risk.

Conversely, should the return of the security be observed to have a lower return (for example, 7%) than the expected return as implied by the SML, the security can be said to be overvalued. The investor receives a reduced return while assuming the same amount of risk.

The Arbitrage Pricing Theory model

The CAPM suffers from several limitations, such as the use of a mean-variance framework and the fact that returns are captured by one risk factor – the market risk factor. In a well-diversified portfolio, the unsystematic risk of various stocks cancels out and is essentially eliminated.

The **Arbitrage Pricing Theory** (**APT**) model was put forward to address these shortcomings and offers a general approach of determining the asset prices other than the mean and variances.

The APT model assumes that the security returns are generated according to multiple factor models, which consist of a linear combination of several systematic risk factors. Such factors could be the inflation rate, GDP growth rate, real interest rates, or dividends.

The equilibrium asset pricing equation according to the APT model is as follows:

$$E[R_i] = \alpha_i + \beta_{i,1} F_1 + \beta_{i,2} F_2 + \ldots + \beta_{i,j} F_j$$

Here, $E[R_i]$ is the expected rate of return on the i security, α_i is the expected return on the i stock if all factors are negligible, $\beta_{i,j}$ is the sensitivity of the ith asset to the jth factor, and F_j is the value of the jth factor that influences the return on the i security.

Since our goal is to find all values of α_i and β, we will perform a **multivariate linear regression** on the APT model.

Multivariate linear regression of factor models

Many Python packages, such as SciPy, come with several variants of regression functions. In particular, the `statsmodels` package is a complement to SciPy with descriptive statistics and the estimation of statistical models. The official page for Statsmodels is `https://www.statsmodels.org`.

If Statsmodels is not yet installed in your Python environment, run the following command to do so:

```
$ pip install -U statsmodels
```

If you have an existing package installed, the -U switch tells `pip` to upgrade the selected package to the newest available version.

In this example, we will use the `ols` function of the `statsmodels` module to perform an ordinary least-squares regression and view its summary.

Let's assume that you have implemented an APT model with seven factors that return the values of Y. Consider the following set of data collected over nine time periods, t_1 to t_9. X_1 to X_7 are independent variables observed at each period. The regression problem is therefore structured as follows:

$$Y = X_{i,1}F_1 + X_{i,2}F_2 + \ldots + X_{i,7}F_7 + c$$

A simple ordinary least-squares regression on values of X and Y can be performed with the following code:

```
In [ ]:
    """
    Least squares regression with statsmodels
    """
    import numpy as np
    import statsmodels.api as sm

    # Generate some sample data
    num_periods = 9
    all_values = np.array([np.random.random(8) \
                            for i in range(num_periods)])

    # Filter the data
    y_values = all_values[:, 0] # First column values as Y
    x_values = all_values[:, 1:] # All other values as X
    x_values = sm.add_constant(x_values) # Include the intercept
    results = sm.OLS(y_values, x_values).fit() # Regress and fit the model
```

Let's view the detailed statistics of the regression:

```
In [ ]:
    print(results.summary())
```

The OLS regression results will output a pretty long table of statistical information. However, our interest lies in one particular section that gives us the coefficients of our APT model:

```
=================================================================
              coef    std err       t      P>|t|     [0.025
-----------------------------------------------------------------
const       0.7229     0.330     2.191    0.273    -3.469
x1          0.4195     0.238     1.766    0.328    -2.599
x2          0.4930     0.176     2.807    0.218    -1.739
x3          0.1495     0.102     1.473    0.380    -1.140
x4         -0.1622     0.191    -0.847    0.552    -2.594
x5         -0.6123     0.172    -3.561    0.174    -2.797
x6         -0.2414     0.161    -1.499    0.375    -2.288
x7         -0.5079     0.200    -2.534    0.239    -3.054
```

The `coef` column gives us the coefficient values of our regression for the c constant, and X_1 until X_7. Similarly, we can use the `params` property to display these coefficients of interest:

```
In [ ]:
    print(results.params)
Out[ ]:
    [ 0.72286627  0.41950411  0.49300959  0.14951292 -0.16218313
 -0.61228465 -0.24143028 -0.50786377]
```

Both the function calls produce the same coefficient values for the APT model in the same order.

Linear optimization

In the CAPM and APT pricing theories, we assumed linearity in the models and solved for expected security prices using regressions in Python.

As the number of securities in our portfolio increases, certain limitations are introduced as well. Portfolio managers would find themselves constrained by these rules in pursuing certain objectives mandated by investors.

Linear optimization helps overcome the problem of portfolio allocation. Optimization focuses on minimizing or maximizing the value of objective functions. Some examples include maximizing returns and minimizing volatility. These objectives are usually governed by certain regulations, such as a no short-selling rule, or limits on the number of securities to be invested.

Unfortunately, in Python, there is no single official package that supports this solution. However, there are third-party packages available with an implementation of the simplex algorithm for linear programming. For the purpose of this demonstration, we will use Pulp, an open source linear programming modeler, to assist us in this particular linear programming problem.

Getting Pulp

You can obtain Pulp from `https://github.com/coin-or/pulp`. The project page contains a comprehensive list of documentation to help you get started with your optimization process.

You may also obtain the Pulp package with the `pip` package manager:

```
$ pip install pulp
```

A maximization example with linear programming

Suppose that we are interested in investing in two securities, X and Y. We would like to find out the actual number of units to invest for every three units of the security X and two units of the security Y, such that the total number of units invested is maximized, where possible. However, there are certain constraints on our investment strategy:

- For every 2 units of the security X invested and 1 unit of the security Y invested, the total volume must not exceed 100
- For every unit of the securities X and Y invested, the total volume must not exceed 80
- The total volume allowed to invest in the security X must not exceed 40
- Short-selling is not allowed for securities

The maximization problem can be mathematically represented as follows:

$$\text{Maximize: } f(x, y) = 3x + 2y$$

subject to:

$$2x + y \leq 100$$

$$x + y \leq 80$$

$$x \leq 40$$

$$x \geq 0, y \geq 0$$

By plotting the constraints on an x by y graph, a set of feasible solutions can be seen, given by the shaded area:

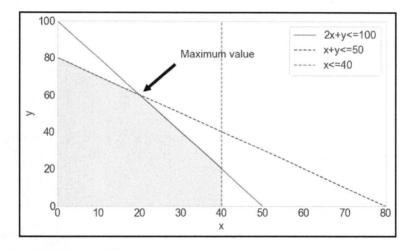

The problem can be translated into Python with the `pulp` package, as follows:

```
In [ ]:
    """
    A simple linear optimization problem with 2 variables
    """
    import pulp

    x = pulp.LpVariable('x', lowBound=0)
    y = pulp.LpVariable('y', lowBound=0)
    problem = pulp.LpProblem(
        'A simple maximization objective',
        pulp.LpMaximize)
    problem += 3*x + 2*y, 'The objective function'
    problem += 2*x + y <= 100, '1st constraint'
    problem += x + y <= 80, '2nd constraint'
```

```
    problem += x <= 40, '3rd constraint'
    problem.solve()
```

The `LpVariable` function declares a variable to be solved. The `LpProblem` function initializes the problem with a text description of the problem and the type of optimization, which in this case is the maximization method. The `+=` operation allows an arbitrary number of constraints to be added, along with a text description. Finally, the `.solve()` method is called to begin performing linear optimization. To show the values solved by the optimizer, use the `.variables()` method to loop through each variable and print out its `varValue`.

The following output is generated when the code runs:

```
In [ ]:
    print("Maximization Results:")
    for variable in problem.variables():
        print(variable.name, '=', variable.varValue)
Out[ ]:
    Maximization Results:
    x = 20.0
    y = 60.0
```

The results show that obtaining the maximum value of 180 is possible when the value of x is 20 and y is 60 while fulfilling the given set of constraints.

Outcomes of linear programs

There are three outcomes in linear optimization, as follows:

- A local optimal solution to a linear program is a feasible solution with a closer objective function value than all other feasible solutions close to it. It may or may not be the **global optimal solution**, a solution that is better than every feasible solution.
- A linear program is **infeasible** if a solution cannot be found.
- A linear program is **unbounded** if the optimal solution is unbounded or is infinite.

Integer programming

In the simple optimization problem we investigated earlier, *A maximization example with linear programming*, the variables were allowed to be continuous or fractional. What if the use of fractional values or results is not realistic? This problem is called the **linear integer programming** problem, where all the variables are restricted as integers. A special case of an integer variable is a binary variable that can either be 0 or 1. Binary variables are especially useful in model decision-making when given a set of choices.

Integer programming models are frequently used in operational research to model real-world working problems. More often than not, stating nonlinear problems in a linear or even binary fashion requires more art than science.

A minimization example with integer programming

Suppose we must go for 150 contracts in a particular over-the-counter exotic security from three dealers. Dealer *X* quoted $500 per contract plus handling fees of $4,000, regardless of the number of contracts sold. Dealer *Y* charges $450 per contract plus a transaction fee of $2,000. Dealer *Z* charges $450 per contract plus a fee of $6,000. Dealer *X* will sell at most 100 contracts, dealer *Y* at most 90, and dealer *Z* at most 70. The minimum transaction volume from any dealer is 30 contracts if any are transacted with that dealer. How should we minimize the cost of purchasing 150 contracts?

Using the `pulp` package, let's set up the required variables:

```
In [ ]:
    """
    An example of implementing an integer
    programming model with binary conditions
    """
    import pulp

    dealers = ['X', 'Y', 'Z']
    variable_costs = {'X': 500, 'Y': 350, 'Z': 450}
    fixed_costs = {'X': 4000, 'Y': 2000, 'Z': 6000}

    # Define PuLP variables to solve
    quantities = pulp.LpVariable.dicts('quantity',
                                       dealers,
                                       lowBound=0,
                                       cat=pulp.LpInteger)
    is_orders = pulp.LpVariable.dicts('orders',
                                      dealers,
                                      cat=pulp.LpBinary)
```

The `dealers` variable simply contains the dictionary identifiers that are used to reference lists and dictionaries later on. The `variable_costs` and `fixed_costs` variables are dictionary objects that contain the respective contract cost and fees charged by each dealer. The Pulp solver solves for the values of `quantities` and `is_orders`, which are defined by the `LpVariable` function. The `dicts()` method tells Pulp to treat the assigned variable as a dictionary object, using the `dealers` variable for referencing. Note that the `quantities` variable has a lower boundary (0) that prevents us from entering a short position in any securities. The `is_orders` values are treated as binary objects, indicating whether we should enter into a transaction with any of the dealers.

What is the best approach to modeling this integer programming problem? At first glance, it seems fairly straightforward by applying this equation:

$$Minimize \sum_{i=x}^{z} isOrder_i(variable\ cost_i \times quantity_i + fixed\ cost_i)$$

Where the following is true:

$$IsOrder_i = \begin{cases} 1, & \text{if buying from dealer } i \\ 0, & \text{if not buying from dealer } i \end{cases}$$

$$30 \leq quantity_x \leq 100$$

$$30 \leq quantity_y \leq 90$$

$$30 \leq quantity_z \leq 70$$

$$\sum_{i=x}^{z} quantity_i = 150$$

The equation simply states that we want to minimize the total costs with the binary variable, *isOrder_i*, to determine whether to account for the costs associated with buying from a specific dealer.

Let's implement this model in Python:

```
In [ ]:
    """
    This is an example of implementing an integer programming model
    with binary variables the wrong way.
    """
    # Initialize the model with constraints
```

```
model = pulp.LpProblem('A cost minimization problem',
                        pulp.LpMinimize)
model += sum([(variable_costs[i] * \
               quantities[i] + \
               fixed_costs[i])*is_orders[i] \
               for i in dealers]), 'Minimize portfolio cost'
model += sum([quantities[i] for i in dealers]) == 150\
        , 'Total contracts required'
model += 30 <= quantities['X'] <= 100\
        , 'Boundary of total volume of X'
model += 30 <= quantities['Y'] <= 90\
        , 'Boundary of total volume of Y'
model += 30 <= quantities['Z'] <= 70\
        , 'Boundary of total volume of Z'
model.solve() # You will get an error running this code!
```

What happens when we run the solver? Check it out:

```
Out [ ]:
    TypeError: Non-constant expressions cannot be multiplied
```

As it turned out, we were trying to perform multiplication on two unknown variables, `quantities` and `is_order`, which unknowingly led us to perform nonlinear programming. Such are the pitfalls encountered when performing integer programming.

How should we solve this problem? We can consider using **binary variables**, as shown in the next section.

Integer programming with binary conditions

Another method for formulating the minimization objective is to place all unknown variables in a linear fashion such that they are additive:

$$Minimize \sum_{i=x}^{z} variable\ cost_i \times quantity_i + fixed\ cost_i \times isOrder_i$$

Comparing with the previous objective equation, we would obtain the same fixed cost values. However, the unknown variable, $quantity_i$, remains in the first term of the equation. Hence, the $quantity_i$ variable is required to be solved as a function of $isOrder_i$, such that the constraints are stated as follows:

$$isOrder_i \times 30 \leq quantity_x \leq isOrder_i \times 100$$

$$isOrder_i \times 30 \leq quantity_y \leq isOrder_i \times 90$$

$$isOrder_i \times 30 \leq quantity_z \leq isOrder_i \times 70$$

Let's apply these formulas in Python:

```
In [ ]:
    """
    This is an example of implementing an
    IP model with binary variables the correct way.
    """
    # Initialize the model with constraints
    model = pulp.LpProblem('A cost minimization problem',
                           pulp.LpMinimize)
    model += sum(
        [variable_costs[i]*quantities[i] + \
            fixed_costs[i]*is_orders[i] for i in dealers])\
        , 'Minimize portfolio cost'
    model += sum([quantities[i] for i in dealers]) == 150\
        ,  'Total contracts required'
    model += is_orders['X']*30 <= quantities['X'] <= \
        is_orders['X']*100, 'Boundary of total volume of X'
    model += is_orders['Y']*30 <= quantities['Y'] <= \
        is_orders['Y']*90, 'Boundary of total volume of Y'
    model += is_orders['Z']*30 <= quantities['Z'] <= \
        is_orders['Z']*70, 'Boundary of total volume of Z'
    model.solve()
```

What happens when we try to run the solver? Let's see:

```
In [ ]:
    print('Minimization Results:')
    for variable in model.variables():
        print(variable, '=', variable.varValue)

    print('Total cost:', pulp.value(model.objective))
Out[ ]:
    Minimization Results:
    orders_X = 0.0
```

```
orders_Y = 1.0
orders_Z = 1.0
quantity_X = 0.0
quantity_Y = 90.0
quantity_Z = 60.0
Total cost: 66500.0
```

The output tells us that buying 90 contracts from dealer *Y* and 60 contracts from dealer *Z* gives us the lowest possible cost of $66,500 while fulfilling all other constraints.

As we can see, careful planning is required in the design of integer programming models to arrive at an accurate solution in order for them to be useful in decision-making.

Solving linear equations using matrices

In the previous section, we looked at solving a system of linear equations with inequality constraints. If a set of systematic linear equations has constraints that are deterministic, we can represent the problem as matrices and apply matrix algebra. Matrix methods represent multiple linear equations in a compact manner while using existing matrix library functions.

Suppose we would like to build a portfolio that consists of three securities: *a*, *b*, and *c*. The allocation of the portfolio must meet certain constraints: it must consist of six units of a long position in the security *a*. With every two units of the security *a*, one unit of the security *b*, and one unit of the security *c* invested, the net position must be long four units. With every one unit of the security *a*, three units of the security *b*, and two units of the security *c* invested, the net position must be long five units.

To find out the number of securities to invest in, we can frame the problem mathematically, as follows:

$$2a + b + c = 4$$

$$a + 3b + 2c = 5$$

$$a = 6$$

With all of the coefficients visible, the equations are as follows:

$$2a + 1b + 1c = 4$$

$$1a + 3b + 2c = 5$$

$$1a + 0b + 0c = 6$$

Let's take the coefficients of the equations and represent them in a matrix form:

$$A = \begin{bmatrix} 2 & 1 & 1 \\ 1 & 3 & 2 \\ 1 & 0 & 0 \end{bmatrix}, x = \begin{bmatrix} a \\ b \\ c \end{bmatrix}, B = \begin{bmatrix} 4 \\ 5 \\ 6 \end{bmatrix}$$

The linear equations can now be stated as follows:

$$Ax = B$$

To solve for the x vector that contains the number of securities to invest in, the inverse of the A matrix is taken and the equation is written as follows:

$$x = A^{-1}B$$

Using NumPy arrays, the A and B matrices are assigned as follows:

```
In [ ]:
    """
    Linear algebra with NumPy matrices
    """
    import numpy as np

    A = np.array([[2, 1, 1],[1, 3, 2],[1, 0, 0]])
    B = np.array([4, 5, 6])
```

We can use the `linalg.solve` function of NumPy to solve a system of linear scalar equations:

```
In [ ]:
    print(np.linalg.solve(A, B))
Out[ ]:
    [  6.   15.  -23.]
```

The portfolio would require a long position of 6 units of the security a, 15 units of the security b, and a short position of 23 units of the security c.

In portfolio management, we can use the matrix system of equations to solve for optimal weight allocations of securities, given a set of constraints. As the number of securities in the portfolio increases, the size of the A matrix increases and it becomes computationally expensive to compute the matrix inversion of A. Thus, one may consider methods such as the Cholesky decomposition, LU decomposition, QR decomposition, the Jacobi method, or the Gauss-Seidel method to break down the A matrix into simpler matrices for factorization.

The LU decomposition

The **LU decomposition**, or also known as **lower-upper factorization**, is one of the methods that solve square systems of linear equations. As its name implies, the LU factorization decomposes the A matrix into a product of two matrices: a lower triangular matrix, L, and an upper triangular matrix, U. The decomposition can be represented as follows:

$$A = LU$$

$$\begin{bmatrix} a & b & c \\ d & e & f \\ g & h & i \end{bmatrix} = \begin{bmatrix} l_{11} & 0 & 0 \\ l_{21} & l_{22} & 0 \\ l_{31} & l_{32} & l_{33} \end{bmatrix} \times \begin{bmatrix} u_{11} & u_{12} & u_{13} \\ 0 & u_{22} & u_{23} \\ 0 & 0 & u_{33} \end{bmatrix}$$

Here, we can see $a=l_{11}u_{11}$, $b=l_{11}u_{12}$, and so on. A lower triangular matrix is a matrix that contains values in its lower triangle with the remaining upper triangle populated with zeros. The converse is true for an upper triangular matrix.

The definite advantage of the LU decomposition method over the Cholesky decomposition method is that it works for any square matrices. The latter only works for symmetric and positive definite matrices.

Think back to the previous example in *Solving linear equations using matrices* of a 3 x 3 A matrix. This time, we will use the `linalg` package of the SciPy module to perform the LU decomposition with the following code:

```
In  [ ]:
    """
    LU decomposition with SciPy
    """
    import numpy as np
    import scipy.linalg as linalg


    # Define A and B
```

```
A = np.array([
    [2., 1., 1.],
    [1., 3., 2.],
    [1., 0., 0.]])
B = np.array([4., 5., 6.])

# Perform LU decomposition
LU = linalg.lu_factor(A)
x = linalg.lu_solve(LU, B)
```

To view the values of x, execute the following command:

```
In  [ ]:
    print(x)
Out[ ]:
    [  6.   15.  -23.]
```

We get the same values of 6, 15, and −23 for *a*, *b*, and *c*, respectively.

Note that we used the `lu_factor()` method of `scipy.linalg` here, which gives the LU variable as the pivoted LU decomposition of the *A* matrix. We used the `lu_solve()` method, which takes in the pivoted LU decomposition and the B vector, to solve the equation system.

We can display the LU decomposition of the A matrix using the `lu()` method of `scipy.linalg`. The `lu()` method returns three variables—the permutation matrix, *P*, the lower triangular matrix, *L*, and the upper triangular matrix, *U* – individually:

```
In [ ]:
    import scipy

    P, L, U = scipy.linalg.lu(A)

    print('P=\n', P)
    print('L=\n', L)
    print('U=\n', U)
```

When we print out these variables, we can conclude the relationship between the LU factorization and *A* matrix, as follows:

$$A = \begin{bmatrix} 2 & 1 & 1 \\ 1 & 3 & 2 \\ 1 & 0 & 0 \end{bmatrix} = \begin{bmatrix} 1 & 0 & 0 \\ 0.5 & 1 & 0 \\ 0.5 & -0.2 & 1 \end{bmatrix} \times \begin{bmatrix} 2 & 1 & 1 \\ 0 & 2.5 & 1.5 \\ 0 & 0 & -0.2 \end{bmatrix}$$

The LU decomposition can be viewed as the matrix form of Gaussian elimination performed on two simpler matrices: the upper triangular and lower triangular matrices.

The Cholesky decomposition

The Cholesky decomposition is another way of solving systems of linear equations. It can be significantly faster and uses a lot less memory than the LU decomposition, by exploiting the property of symmetric matrices. However, the matrix being decomposed must be Hermitian (or real-valued symmetric and thus square) and positive definite. This means that the A matrix is decomposed as $A=LL^T$, where L is a lower triangular matrix with real and positive numbers on the diagonals, and L^T is the conjugate transpose of L.

Let's consider another example of a system of linear equations where the A matrix is both Hermitian and positive definite. Again, the equation is in the form of $Ax=B$, where A and B take the following values:

$$A = \begin{bmatrix} 10 & -1 & 2 & 0 \\ -1 & 11 & -1 & 3 \\ 2 & -1 & 10 & -1 \\ 0 & 3 & -1 & 8 \end{bmatrix}, x = \begin{bmatrix} a \\ b \\ c \\ d \end{bmatrix}, B = \begin{bmatrix} 6 \\ 25 \\ -11 \\ 15 \end{bmatrix}$$

Let's represent these matrices as NumPy arrays:

```
In  [ ]:
    """
    Cholesky decomposition with NumPy
    """
    import numpy as np

    A = np.array([
        [10., -1., 2., 0.],
        [-1., 11., -1., 3.],
        [2., -1., 10., -1.],
        [0., 3., -1., 8.]])
    B = np.array([6., 25., -11., 15.])

    L = np.linalg.cholesky(A)
```

The `cholesky()` function of `numpy.linalg` would compute the lower triangular factor of the A matrix. Let's view the lower triangular matrix:

```
In  [ ]:
    print(L)
Out[ ]:
    [[ 3.16227766  0.          0.          0.        ]
     [-0.31622777  3.3015148   0.          0.        ]
     [ 0.63245553 -0.24231301  3.08889696  0.        ]
     [ 0.          0.9086738  -0.25245792  2.6665665 ]]
```

To verify that the Cholesky decomposition results are correct, we can use the definition of the Cholesky factorization by multiplying L by its conjugate transpose, which will lead us back to the values of the A matrix:

```
In  [ ]:
    print(np.dot(L, L.T.conj()))  # A=L.L*
Out [ ]:
    [[10. -1.   2.   0.]
     [-1. 11.  -1.   3.]
     [ 2. -1.  10.  -1.]
     [ 0.  3.  -1.   8.]]
```

Before solving for x, we need to solve for $L^T x$ as y. Let's use the `solve()` method of `numpy.linalg`:

```
In  [ ]:
    y = np.linalg.solve(L, B)   # L.L*.x=B; When L*.x=y, then L.y=B
```

To solve for x, we need to solve again using the conjugate transpose of L and y:

```
In  [ ]:
    x = np.linalg.solve(L.T.conj(), y)   # x=L*'.y
```

Let's print our result of x:

```
In  [ ]:
    print(x)
Out[ ]:
    [ 1.  2. -1.  1.]
```

The output gives us our values of x for a, b, c, and d.

To show that the Cholesky factorization gives us the correct values, we can verify the answer by multiplying the A matrix by the transpose of x to return the values of B:

```
In [ ] :
    print(np.mat(A) * np.mat(x).T)   # B=Ax
Out[ ]:
    [[  6.]
     [ 25.]
     [-11.]
     [ 15.]]
```

This shows that the values of x by the Cholesky decomposition would lead to the same values given by B.

The QR decomposition

The **QR decomposition**, also known as the **QR factorization**, is another method of solving linear systems of equations using matrices, very much like the LU decomposition. The equation to solve is in the form of *Ax=B*, where matrix *A=QR*. However, in this case, *A* is a product of an orthogonal matrix, *Q*, and upper triangular matrix, *R*. The QR algorithm is commonly used to solve the linear least-squares problem.

An orthogonal matrix exhibits the following properties:

- It is a square matrix.
- Multiplying an orthogonal matrix by its transpose returns the identity matrix:

$$QQ^T = Q^TQ = 1$$

- The inverse of an orthogonal matrix equals its transpose:

$$Q^T = Q^{-1}$$

An identity matrix is also a square matrix, with its main diagonal containing 1s and 0s elsewhere.

The problem of *Ax=B* can now be restated as follows:

$$QRx = B$$

$$Rx = Q^{-1}B \ or \ Rx = Q^TB$$

Using the same variables in the LU decomposition example, we will use the `qr()` method of `scipy.linalg` to compute our values of *Q* and *R*, and let the *y* variable represent our value of BQ^T with the following code:

```
In  [ ]:
    """
    QR decomposition with scipy
    """
    import numpy as np
    import scipy.linalg as linalg


    A = np.array([
        [2., 1., 1.],
        [1., 3., 2.],
        [1., 0., 0]])
```

```
    B = np.array([4., 5., 6.])

    Q, R = scipy.linalg.qr(A)   # QR decomposition
    y = np.dot(Q.T, B)   # Let y=Q'.B
    x = scipy.linalg.solve(R, y)   # Solve Rx=y
```

Note that Q.T is simply the transpose of Q, which is also the same as the inverse of *Q*:

```
In [ ]:
    print(x)
Out[ ]:
    [  6.   15.  -23.]
```

We get the same answers as those in the LU decomposition example.

Solving with other matrix algebra methods

So far, we've looked at the use of matrix inversion, the LU decomposition, the Cholesky decomposition, and QR decomposition to solve for systems of linear equations. Should the size of our financial data in the *A* matrix be large, it can be broken down by a number of schemes so that the solution can converge more quickly using matrix algebra. Quantitative portfolio analysts should be familiar with these methods.

In some circumstances, the solution that we are looking for might not converge. Therefore, you might consider the use of iterative methods. Common methods to solve systems of linear equations iteratively are the Jacobi method, the Gauss-Seidel method, and the SOR method. We will take a brief look at examples of implementing the Jacobi and Gauss-Seidel methods.

The Jacobi method

The Jacobi method solves a system of linear equations iteratively along its diagonal elements. The iteration procedure terminates when the solution converges. Again, the equation to solve is in the form of *Ax=B*, where the matrix *A* can be decomposed into two matrices of the same size such that *A=D+R*. The matrix D consists of only the diagonal components of A, and the other matrix R consists of the remaining components. Let's take a look at the example of a 4 x 4 *A* matrix:

$$A = \begin{bmatrix} a & b & c & d \\ e & f & g & h \\ i & j & k & l \\ m & n & o & p \end{bmatrix} = \begin{bmatrix} a & 0 & 0 & 0 \\ 0 & f & 0 & 0 \\ 0 & 0 & k & 0 \\ 0 & 0 & 0 & p \end{bmatrix} + \begin{bmatrix} 0 & b & c & d \\ e & 0 & g & h \\ i & j & 0 & l \\ m & n & o & 0 \end{bmatrix}$$

The solution is then obtained iteratively, as follows:

$$Ax = B$$

$$(D + R)x = B$$

$$Dx = B - Rx$$

$$X_{n+1} = D^{-1}(B - Rx_n)$$

As opposed to the Gauss-Siedel method, the value of x_n in the Jacobi method is needed during each iteration in order to compute x_{n+1} and cannot be overwritten. This would take up twice the amount of storage. However, the computations for each element can be done in parallel, which is useful for faster computations.

If the A matrix is strictly irreducibly diagonally dominant, this method is guaranteed to converge. A strictly irreducibly diagonally dominant matrix is one where the absolute diagonal element in every row is greater than the sum of the absolute values of other terms.

In some circumstances, the Jacobi method can converge even if these conditions are not met. The Python code is given as follows:

```
In [ ]:
    """
    Solve Ax=B with the Jacobi method
    """
    import numpy as np

    def jacobi(A, B, n, tol=1e-10):
        # Initializes x with zeroes with same shape and type as B
        x = np.zeros_like(B)

        for iter_count in range(n):
            x_new = np.zeros_like(x)
            for i in range(A.shape[0]):
                s1 = np.dot(A[i, :i], x[:i])
                s2 = np.dot(A[i, i + 1:], x[i + 1:])
                x_new[i] = (B[i] - s1 - s2) / A[i, i]

            if np.allclose(x, x_new, tol):
                break

            x = x_new

        return x
```

Consider the same matrix values in the Cholesky decomposition example. We will use 25 iterations in our `jacobi` function to find the values of x:

```
In [ ] :
    A = np.array([
        [10., -1., 2., 0.],
        [-1., 11., -1., 3.],
        [2., -1., 10., -1.],
        [0.0, 3., -1., 8.]])
    B = np.array([6., 25., -11., 15.])
    n = 25
```

After initializing the values, we can now call the function and solve for x:

```
In [ ]:
    x = jacobi(A, B, n)
    print('x', '=', x)
Out[ ]:
    x = [ 1.  2. -1.  1.]
```

We solved for the values of x, which are similar to the answers from the Cholesky decomposition.

The Gauss-Seidel method

The Gauss-Seidel method works very much like the Jacobi method. It is another way to solve a square system of linear equations using an iterative procedure with the equation in the form of $Ax=B$. Here, the A matrix is decomposed as $A=L+U$, where the A matrix is a sum of a lower triangular matrix, L, and an upper triangular matrix, U. Let's take a look at the example of a 4 x 4 A matrix:

$$A = \begin{bmatrix} a & b & c & d \\ e & f & g & h \\ i & j & k & l \\ m & n & o & p \end{bmatrix} = \begin{bmatrix} a & 0 & 0 & 0 \\ e & f & 0 & 0 \\ i & j & k & 0 \\ m & n & o & p \end{bmatrix} + \begin{bmatrix} 0 & b & c & d \\ 0 & 0 & g & h \\ 0 & 0 & 0 & l \\ 0 & 0 & 0 & 0 \end{bmatrix}$$

The solution is then obtained iteratively, as follows:

$$Ax = B$$

$$(L + U)x = B$$

$$Lx = B - Ux$$

$$X_{n+1} = L^{-1}(B - Ux_n)$$

Using a lower triangular matrix, L, where zeroes fill up the upper triangle, the elements of x_n can be overwritten in each iteration in order to compute x_{n+1}. This results in the advantage of needing half the storage required when using the Jacobi method.

The rate of convergence in the Gauss-Seidel method largely lies in the properties of the A matrix, especially if the A matrix is needed to be strictly-diagonally dominant or symmetric positive definite. Even if these conditions are not met, the Gauss-Seidel method may converge.

The Python implementation of the Gauss-Seidel method is given as follows:

```
In  [ ]:
    """
    Solve Ax=B with the Gauss-Seidel method
    """
    import numpy as np

    def gauss(A, B, n, tol=1e-10):
        L = np.tril(A)   # returns the lower triangular matrix of A
        U = A-L   # decompose A = L + U
        L_inv = np.linalg.inv(L)
        x = np.zeros_like(B)

        for i in range(n):
            Ux = np.dot(U, x)
            x_new = np.dot(L_inv, B - Ux)

            if np.allclose(x, x_new, tol):
                break

            x = x_new

        return x
```

Here, the `tril()` method of NumPy returns the lower triangular A matrix, from which we can find the lower triangular U matrix. Plugging the remaining values into x iteratively would lead us to the following solution, with some tolerance defined by `tol`.

Let's consider the same matrix values in the Jacobi method and Cholesky decomposition example. We will use a maximum of 100 iterations in our `gauss()` function to find the values of x, as follows:

```
In  [ ]:
    A = np.array([
        [10., -1., 2., 0.],
        [-1., 11., -1., 3.],
        [2., -1., 10., -1.],
        [0.0, 3., -1., 8.]])
    B = np.array([6., 25., -11., 15.])
    n = 100
    x = gauss(A, B, n)
```

Let's see whether our x values match with those from the Jacobi method and Cholesky decomposition:

```
In [ ]:
    print('x', '=', x)
Out[ ]:
    x = [ 1.  2. -1.  1.]
```

We solved for the values of x, which are similar to the answers from the Jacobi method and Cholesky decomposition.

Summary

In this chapter, we took a brief look at the use of the CAPM model and APT model in finance. In the CAPM model, we visited the efficient frontier with the CML to determine the optimal portfolio and the market portfolio. Then, we solved for the SML using regression, which helped us to determine whether an asset is undervalued or overvalued. In the APT model, we explored how various factors affect security returns other than using the mean-variance framework. We performed a multivariate linear regression to help us determine the coefficients of the factors that led to the valuation of our security price.

In portfolio allocation, portfolio managers are typically mandated by investors to achieve a set of objectives while following certain constraints. We can model this problem using linear programming. Using the Pulp Python package, we can define a minimization or maximization objective function, and add inequality constraints to our problems to solve for unknown variables. The three outcomes in linear optimization can be an unbounded solution, only one solution, or no solution at all.

Another form of linear optimization is integer programming, where all the variables are restricted to being integers instead of fractional values. A special case of an integer variable is a binary variable, which can either be 0 or 1, and it is especially useful to model decision-making when given a set of choices. We worked on a simple integer programming model that contains binary conditions and saw how easy it is to run into a pitfall. Careful planning on the design of integer programming models is required for them to be useful in decision-making.

The portfolio-allocation problem may also be represented as a system of linear equations with equalities, which can be solved using matrices in the form of $Ax=B$. To find the values of x, we solved for $A^{-1}B$ using various types of decomposition of the A matrix. The two types of matrix decomposition method are the direct and indirect methods. The direct method performs matrix algebra in a fixed number of iterations, and includes the LU decomposition, Cholesky decomposition, and QR decomposition methods. The indirect or iterative method iteratively computes the next values of x until a certain tolerance of accuracy is reached. This method is particularly useful for computing large matrices, but it also faces the risk of not having the solution converge. The indirect methods we used are the Jacobi method and the Gauss-Seidel method.

In the next chapter, we will look at nonlinear modeling in finance.

3
Nonlinearity in Finance

In recent years, there has been a growing interest in research on nonlinear phenomena in economic and financial theory. With nonlinear serial dependence playing a significant role in the returns of many financial time series, this makes security valuation and pricing very important, leading to an increase in studies on the nonlinear modeling of financial products.

Practitioners in the financial industry use nonlinear models to forecast volatility, price derivatives, and compute **Value at Risk (VAR)**. Unlike linear models, where linear algebra is used to find a solution, nonlinear models do not necessarily infer a global optimal solution. Numerical root-finding methods are usually employed to converge toward the nearest local optimal solution, which is a root.

In this chapter, we will discuss the following topics:

- Nonlinearity modeling
- Examples of nonlinear models
- Root-finding algorithms
- SciPy implementations in root-finding

Nonlinearity modeling

While linear relationships aim to explain observed phenomena in the simplest way possible, many complex physical phenomena cannot be explained using such models. A nonlinear relationship is defined as follows:

$$f(a + b) \neq f(a) + f(b)$$

Even though nonlinear relationships may be complex, to fully understand and model them, we will take a look at the examples that are applied in the context of finance and in time-series models.

Examples of nonlinear models

Many nonlinear models have been proposed for academic and applied research to explain certain aspects of economic and financial data that are left unexplained by linear models. The literature on nonlinearity in finance is simply too broad and deep to be adequately explained in this book. In this section, we will briefly discuss some examples of nonlinear models that we may come across for practical uses: the implied volatility model, Markov switching model, threshold model, and smooth transition model.

The implied volatility model

Perhaps one of the most widely-studied option-pricing models is the Black-Scholes-Merton model, or simply the Black-Scholes model. A call option is a right, but not an obligation, to buy the underlying security at a particular price and time. A put option is a right, but not an obligation, to sell the underlying security at a particular price and time. The Black-Scholes model helps determine the fair price of an option with the assumption that the returns of the underlying security are normally distributed (N(.)) or that asset prices are log-normally distributed.

The formula takes on the following assumed variables—the strike price (K), the time to expiry (T), the risk-free rate (r), the volatility of the underlying returns (σ), the current price of the underlying asset (S), and its yield (q). The mathematical formula for a call option, $C(S,t)$, is represented as follows:

$$C(S,t) = Se^{qT}N(d_1) - Ke^{-rT}N(d_2)$$

Here:

$$d_1 = \frac{ln(S/K) + (r - q + \sigma^2/2)T}{\sigma\sqrt{T}}$$

By way of market forces, the price of an option may deviate from the price that's been derived from the Black-Scholes formula. In particular, the realized volatility (that is, the observed volatility of the underlying returns from historical market prices) could differ from the volatility value as implied by the Black-Scholes model, which is indicated by σ.

Think back to the **Capital Asset Pricing Model** (**CAPM**) discussed in `Chapter 2`, *The Importance of Linearity in Finance*. In general, securities that have higher returns exhibit higher risk, as indicated by the volatility or standard deviation of returns.

With volatility being such an important factor in security pricing, many volatility models have been proposed for studies. One such model is the implied volatility modeling of option prices.

Suppose we plot the implied volatility values of an equity option given by the Black-Scholes formula with a particular maturity for every strike price available. In general, we get a curve commonly known as the **volatility smile** due to its shape:

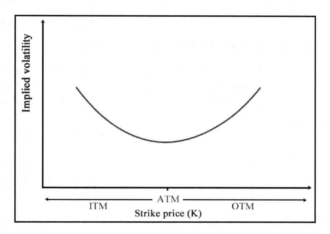

The **Implied volatility** typically is at its highest for deep in-the-money and out-of-the-money options driven by heavy speculation and at its lowest for at-the-money options.

The characteristics of options are explained as follows:

- **In-the-money options** (**ITM**): A call option is considered ITM when its strike price is below the market price of the underlying asset. A put option is considered ITM when its strike price is above the market price of the underlying asset. ITM options have an intrinsic value when exercised.
- **Out-of-the-money options** (**OTM**): A call option is considered OTM when its strike price is above the market price of the underlying asset. A put option is considered OTM when its strike price is below the market price of the underlying asset. OTM options have no intrinsic value when exercised, but may still have time value.
- **At-the-money options** (**ATM**): An option is considered ATM when its strike price is the same as the market price of the underlying asset. ATM options have no intrinsic value, but may still have time value.

From the preceding volatility curve, one of the objectives in implied volatility modeling is to seek the lowest implied volatility value possible or, in other words, to *find the root*. When found, the theoretical price of an ATM option for a particular maturity can be deduced and compared against the market prices for potential opportunities, such as for studying near ATM options or far OTM options. However, since the curve is nonlinear, linear algebra cannot adequately solve the root. We will take a look at a number of root-finding methods in the next section, *Root-finding algorithms*.

The Markov regime-switching model

To model nonlinear behavior in economic and financial time series, Markov switching models can be used to characterize time series in different states of the world or regimes. Examples of such states could be a *volatile* state, as seen in the 2008 global economic downturn, or the *growth* state of a steadily recovering economy. The ability to transition between these structures lets the model capture complex dynamic patterns.

The Markov property of stock prices implies that only the present values are relevant for predicting the future. Past stock-price movements are irrelevant to the way the present has emerged.

Let's take an example of a Markov regime-switching model with *m=2* regimes:

$$y_t = \begin{cases} x_1 + \epsilon_t, \text{when } s_t = 1 \\ x_2 + \epsilon_t, \text{when } s_t = 2 \end{cases}$$

ϵ_t is an **independent and identically-distributed (i.i.d)** white noise. White noise is a normally-distributed random process with a mean of zero. The same model can be represented with dummy variables:

$$y_t = x_1 D_t + x_2 (1 - D_t) + \epsilon_t$$

$$\text{where } D_t = 1 \text{ when } s_t = 1$$

$$\text{or } D_t = 0 \text{ when } s_t = 2$$

The application of Markov switching models includes representing the real GDP growth rate and inflation rate dynamics. These models in turn drive the valuation models of interest-rate derivatives. The probability of switching from the previous state, *i*, to the current state, *j*, can be written as follows:

$$P[s_t = j | s_{t-1} = i]$$

The threshold autoregressive model

One popular class of nonlinear time series models is the **threshold autoregressive** (**TAR**) model, which looks very similar to the Markov switching models. Using regression methods, simple AR models are arguably the most popular models to explain nonlinear behavior. Regimes in the threshold model are determined by past, d, values of its own time series, relative to a threshold value, c.

The following is an example of a **self-exciting TAR** (**SETAR**) model. The SETAR model is self-exciting because switching between different regimes depends on the past values of its own time series:

$$y_t = \begin{cases} a_1 + b_1 y_{t-d} + \epsilon_t, \text{if } y_{t-d} \leq c \\ a_2 + b_2 y_{t-d} + \epsilon_t, \text{if } y_{t-d} > c \end{cases}$$

Using dummy variables, the SETAR model can also be represented as follows:

$$y_t = (a_1 + b_1 y_{t-d})D_t + (a_2 + b_2 y_{t-d})(1 - D_t) + \epsilon_t$$

$$\text{where } D_t = 1 \text{ when } y_{t-d} \leq c,$$

$$\text{or } D_t = 0 \text{ when } y_{t-d} > c$$

 The use of the TAR model may result in sharp transitions between the states as controlled by the threshold variable, c.

Smooth transition models

Abrupt regime changes in the threshold models appear to be unrealistic against real-world dynamics. This problem can be overcome by introducing a smoothly-changing continuous function from one regime to another. The SETAR model becomes a **logistic smooth transition threshold autoregressive** (**LSTAR**) model with the logistic function of $G(y_{t-1};\gamma,c)$:

$$G(y_{t-1}; \gamma, c) = \frac{1}{1 + e^{-\gamma(y_{t-d} - c)}}$$

The SETAR model now becomes an LSTAR model, as shown in the following equation:

$$y_t = (a_1 + b_1 y_{t-d})(1 - G(y_{t-1}; \gamma, c)) + (a_2 + b_2 y_{t-d})G(y_{t-1}; \gamma, c) + \epsilon_t$$

The γ parameter controls the smooth transition from one regime to another. For large values of γ, the transition is the fastest, as y_{t-d} approaches the threshold variable, c. When $\gamma=0$, the LSTAR model is equivalent to a simple *AR(1)* one-regime model.

Root-finding algorithms

In the preceding section, we discussed some nonlinear models commonly used for studying economics and financial time series. From the model data given in continuous time, the intention is therefore to search for the extrema that could possibly infer valuable information. The use of numerical methods, such as root-finding algorithms, can help us find the roots of a continuous function, *f*, such that *f(x)=0*, which can either be the maxima or the minima of the function. In general, an equation may either contain a number of roots or none at all.

One example of the use of root-finding methods on nonlinear models is the Black-Scholes implied volatility modeling discussed earlier, in *The implied volatility model* section. An option trader would be interested in deriving implied prices based on the Black-Scholes model and comparing them with market prices. In the next chapter, `Numerical Methods for Pricing Options`, we will see how we can combine a root-finding method with a numerical-option pricing procedure to create an implied volatility model based on the market prices of a particular option.

Root-finding methods use an iterative routine that requires a start point or the estimation of the root. The estimation of the root can either converge toward a solution, converge to a root that is not sought, or may not even find a solution at all. Thus, it is crucial to find a good approximation to the root.

Not every nonlinear function can be solved using root-finding methods. The following figure shows an example of a continuous function, $\frac{1}{x^2 - 2x}$, where root-finding methods may fail to arrive at a solution. There are discontinuities at *x=0* and *x=2* for the *y* values in the range of -20 to 20:

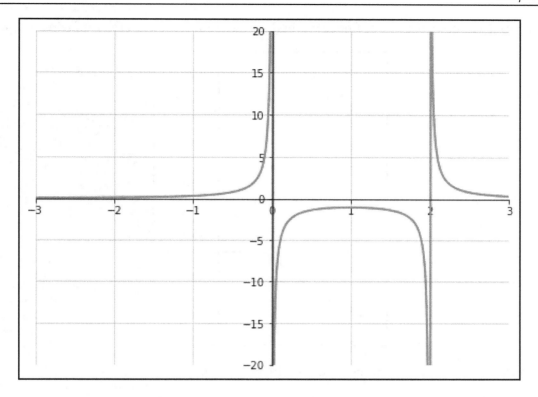

There is no fixed rule as to how a good approximation can be defined. It is recommended that you bracket or define the lower and upper search bounds before starting the root-finding iterative procedure. We certainly do not want to keep searching endlessly in the wrong direction for our root.

Incremental search

A crude method of solving a nonlinear function is by doing an incremental search. Using an arbitrary starting point, a, we can obtain values of $f(a)$ for every increment of dx. We assume that the values of $f(a+dx)$, $f(a+2dx)$, $f(a+3dx)$... are going in the same direction as indicated by their sign. Once the sign changes, a solution is deemed as found. Otherwise, the iterative search terminates when it crosses the boundary point, b.

A pictorial example of the root-finder method for iteration is given in the following graph:

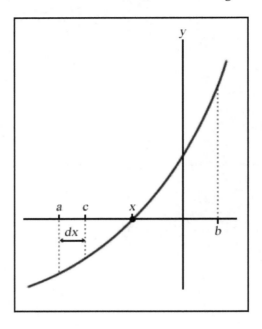

An example can be seen from the following Python code:

```
In [ ]:
    """
    An incremental search algorithm
    """
    import numpy as np

    def incremental_search(func, a, b, dx):
        """
        :param func: The function to solve
        :param a: The left boundary x-axis value
        :param b: The right boundary x-axis value
        :param dx: The incremental value in searching
        :return:
            The x-axis value of the root,
            number of iterations used
        """
        fa = func(a)
        c = a + dx
        fc = func(c)
        n = 1
        while np.sign(fa) == np.sign(fc):
```

```
        if a >= b:
            return a - dx, n

        a = c
        fa = fc
        c = a + dx
        fc = func(c)
        n += 1

    if fa == 0:
        return a, n
    elif fc == 0:
        return c, n
    else:
        return (a + c)/2., n
```

At every iterative procedure, a will be replaced by c, and c will be incremented by dx before the next comparison. Should a root be found, it is plausible that it lies between a and c, both inclusive. In the event that the solution does not rest at either point, we will simply return the average of the two points as the best estimation. The *n* variable keeps track of the number of iterations that underwent the process of finding our root.

We will use the equation that has an analytic solution of $y = x^3 + 2x^2 - 5$ to demonstrate and measure our root-finder, where *x* is bounded between -5 and 5. A small *dx* value of 0.001 is given, which also acts as a precision tool. Smaller values of *dx* produce better precision but also require more search iterations:

```
In [ ]:
    # The keyword 'lambda' creates an anonymous function
    # with input argument x
    y = lambda x: x**3 + 2.*x**2 - 5.
    root, iterations = incremental_search (y, -5., 5., 0.001)
    print("Root is:", root)
    print("Iterations:", iterations)
Out [ ]:
    Root is: 1.2414999999999783
    Iterations: 6242
```

The incremental search root-finder method is a basic demonstration of the fundamental behavior of a root-finding algorithm. The accuracy is at its best when defined by *dx*, and consumes an extremely long computational time in the worst possible scenario. The higher the accuracy demanded, the longer it takes for the solution to converge. For practical reasons, this method is the least preferred of all root-finding algorithms, and we will take a look at alternative methods to find the roots of our equation that can give us better performance.

The bisection method

The bisection method is considered the simplest one-dimensional root-finding algorithm. The general interest is to find the value, x, of a continuous function, f, such that $f(x)=0$.

Suppose we know the two points of an interval, a and b, where $a < b$, and that $f(a)<0$ and $f(b)>0$ lie along the continuous function, taking the midpoint of this interval as c, where $c = \dfrac{a+b}{2}$; the bisection method then evaluates this value as $f(c)$.

Let's illustrate the setup of points along a nonlinear function with the following graph:

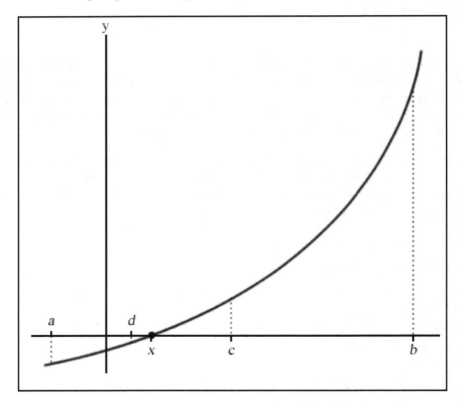

Since the value of $f(a)$ is negative and $f(b)$ is positive, the bisection method assumes that the root, x, lies somewhere between a and b and gives $f(x)=0$.

If $f(c)=0$ or is very close to zero by some predetermined error-tolerance value, a root is declared as found. If $f(c)<0$, we may conclude that a root exists along the c and b interval, or the a and c interval otherwise.

In the next evaluation, *c* is replaced as either *a* or *b* accordingly. With the new interval shortened, the bisection method repeats with the same evaluation to determine the next value of *c*. This process continues, shrinking the width of the *ab* interval until the root is considered found.

The biggest advantage of using the bisection method is its guarantee to converge to an approximation of the root, given a predetermined error-tolerance level and the maximum number of iterations allowed. It should be noted that the bisection method does not require knowledge of the derivative of the unknown function. In certain continuous functions, the derivative could be complex or even impossible to calculate. This makes the bisection method extremely valuable for working on functions that are not smooth.

Because the bisection method does not require derivative information from the continuous function, its major drawback is that it takes up more computational time in the iterative evaluation than other root-finder methods. Also, since the search boundary of the bisection method lies in the *a* and *b* intervals, it would require a good approximation to ensure that the root falls within this range. Otherwise, an incorrect solution may be obtained, or even none at all. Using large values of *a* and *b* might consume more computational time.

The bisection is considered to be stable without the use of an initial guess value for convergence to happen. Often, it is used in combination with other methods, such as the faster Newton's method, to converge quickly with precision.

The Python code for the bisection method is given as follows. Save this as `bisection.py`:

```
In [ ]:
    """
    The bisection method
    """
    def bisection(func, a, b, tol=0.1, maxiter=10):
        """
        :param func: The function to solve
        :param a: The x-axis value where f(a)<0
        :param b: The x-axis value where f(b)>0
        :param tol: The precision of the solution
        :param maxiter: Maximum number of iterations
        :return:
            The x-axis value of the root,
            number of iterations used
        """
        c = (a+b)*0.5  # Declare c as the midpoint ab
        n = 1  # Start with 1 iteration
        while n <= maxiter:
            c = (a+b)*0.5
            if func(c) == 0 or abs(a-b)*0.5 < tol:
```

```
                    # Root is found or is very close
                    return c, n

            n += 1
            if func(c) < 0:
                a = c
            else:
                b = c

        return c, n
In [ ]:
    y = lambda x: x**3 + 2.*x**2 - 5
    root, iterations = bisection(y, -5, 5, 0.00001, 100)
    print("Root is:", root)
    print("Iterations:", iterations)
Out[ ]:
    Root is: 1.241903305053711
    Iterations: 20
```

Again, we bounded the anonymous `lambda` function to the y variable with an input parameter, x, and attempted to solve the $y = x^3 + 2x^2 - 5$ equation as before, in the interval between -5 to 5 to an accuracy of 0.00001 with a maximum iteration of 100.

As we can see, the result from the bisection method gives us better precision in far fewer iterations over the incremental search method.

Newton's method

Newton's method, also known as the **Newton-Raphson method**, uses an iterative procedure to solve for a root using information about the derivative of a function. The derivative is treated as a linear problem to be solved. The first-order derivation, f', of the function, f, represents the tangent line. The approximation to the next value of x, given as x_1, is as follows:

$$x_1 = x - \frac{f(x)}{f'(x)}$$

Here, the tangent line intersects the x axis at x_1, which produces $y=0$. This also represents the first-order Taylor expansion about x_1, such that that the new point, $x_1 = x + \Delta x$, solves the following equation:

$$f(x_1 + \Delta x) = 0$$

This process is repeated with x taking the value of x_1 until the maximum number of iterations is reached, or the absolute difference between x_1 and x is within an acceptable accuracy level.

An initial guess value is required to compute the values of $f(x)$ and $f'(x)$. The rate of convergence is quadratic, which is considered to be extremely fast at obtaining the solution with high levels of accuracy.

The drawback to Newton's method is that it does not guarantee global convergence to the solution. Such a situation arises when the function contains more than one root, or when the algorithm arrives at a local extremum and is unable to compute the next step. As this method requires knowledge of the derivative of its input function, it is required that the input function be differentiable. However, in certain circumstances, it is impossible for the derivative of a function to be known, or otherwise be mathematically easy to compute.

A graphical representation of Newton's method is shown in the following screenshot. x_0 is the initial x value. The derivative of $f(x_0)$ is evaluated, which is a tangent line crossing the x axis at x_1. The iteration is repeated, evaluating the derivative at points x_1, x_2, x_3, and so on:

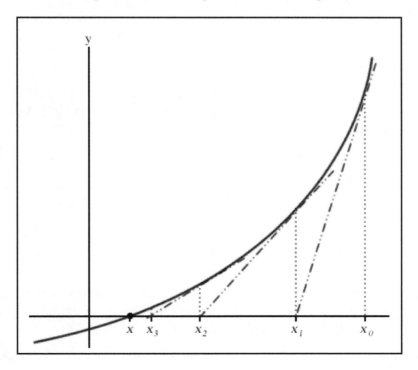

The implementation of Newton's method in Python is as follows:

```
In  [ ]:
    """
    The Newton-Raphson method
    """
    def newton(func, df, x, tol=0.001, maxiter=100):
        """
        :param func: The function to solve
        :param df: The derivative function of f
        :param x: Initial guess value of x
        :param tol: The precision of the solution
        :param maxiter: Maximum number of iterations
        :return:
            The x-axis value of the root,
            number of iterations used
        """
        n = 1
        while n <= maxiter:
            x1 = x - func(x)/df(x)
            if abs(x1 - x) < tol: # Root is very close
                return x1, n

            x = x1
            n += 1

        return None, n
```

We will use the same function used in the bisection example and take a look at the results from Newton's method:

```
In  [ ]:
    y = lambda x: x**3 + 2*x**2 - 5
    dy = lambda x: 3*x**2 + 4*x
    root, iterations = newton(y, dy, 5.0, 0.00001, 100)
    print("Root is:", root)
    print("Iterations:", iterations)
Out [ ]:
    Root is: 1.241896563034502
    Iterations: 7
```

Beware of division by zero exceptions! In Python 2, using values such as 5.0, instead of 5, lets Python recognize the variable as a float, avoids the problem of treating variables as integers in calculations, and gives us better precision.

With Newton's method, we obtained a really close solution with less iteration over the bisection method.

The secant method

The secant method uses secant lines to find the root. A secant line is a straight line that intersects two points of a curve. In the secant method, a line is drawn between two points on the continuous function, such that it extends and intersects the x axis. This method can be thought of as a Quasi-Newton method. By successively drawing such secant lines, the root of the function can be approximated.

The secant method is graphically represented in the following screenshot. An initial guess of the two x axis values, a and b, is required to find $f(a)$ and $f(b)$. A secant line, y, is drawn from $f(b)$ to $f(a)$ and intersects at the c point on the x axis, such that:

$$y = \frac{f(b) - f(a)}{b - a}(c - b) + f(b)$$

The solution to c is therefore as follows:

$$c = b - f(b)\frac{b - a}{f(b) - f(a)}$$

On the next iteration, a and b will take on the values of b and c, respectively. The method repeats itself, drawing secant lines for the x axis values of a and b, b and c, c and d, and so on. The solution terminates when the maximum number of iterations has been reached, or the difference between b and c has reached a pre-specified tolerance level, as shown in the following graph:

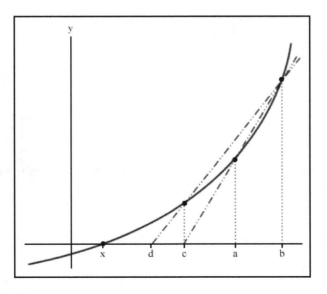

The rate of convergence of the secant method is considered to be superlinear. Its secant method converges much faster than the bisection method and slower than Newton's method. In Newton's method, the number of floating-point operations takes up twice as much time as the secant method in the computation of both its function and its derivative on every iteration. Since the secant method requires only the computation of its function at every step, it can be considered faster in absolute time.

The initial guess values of the secant method must be close to the root, otherwise it has no guarantee of converging to the solution.

The Python code for the secant method is given as follows:

```
In [ ]:
    """
    The secant root-finding method
    """
    def secant(func, a, b, tol=0.001, maxiter=100):
        """
        :param func: The function to solve
        :param a: Initial x-axis guess value
        :param b: Initial x-axis guess value, where b>a
        :param tol: The precision of the solution
        :param maxiter: Maximum number of iterations
        :return:
            The x-axis value of the root,
            number of iterations used
        """
        n = 1
        while n <= maxiter:
            c = b - func(b)*((b-a)/(func(b)-func(a)))
            if abs(c-b) < tol:
                return c, n

            a = b
            b = c
            n += 1

        return None, n
```

Again, we will reuse the same nonlinear function and return the results from the secant method:

```
In [ ]:
    y = lambda x: x**3 + 2.*x**2 - 5.
    root, iterations = secant(y, -5.0, 5.0, 0.00001, 100)
    print("Root is:", root)
    print("Iterations:", iterations)
```

```
Out[ ]:
    Root is: 1.2418965622558549
    Iterations: 14
```

Though all of the preceding root-finding methods gave very close solutions, the secant method performs with fewer iterations compared to the bisection method, but with more than Newton's method.

Combing root-finding methods

It is perfectly possible to write your own root-finding algorithms using a combination of the previously-mentioned root-finding methods. For example, you may use the following implementation:

1. Use the faster secant method to converge the problem to a pre-specified error-tolerance value or a maximum number of iterations
2. Once a pre-specified tolerance level is reached, switch to using the bisection method to converge to the root by halving the search interval with each iteration until the root is found

Brent's method or the **Wijngaarden-Dekker-Brent method** combines the bisection root-finding method, secant method, and inverse quadratic interpolation. The algorithm attempts to use either the secant method or inverse quadratic interpolation whenever possible, and uses the bisection method where necessary. Brent's method can also be found in the `scipy.optimize.brentq` function of SciPy.

SciPy implementations in root-finding

Before starting to write your root-finding algorithm to solve nonlinear or even linear problems, take a look at the documentation of the `scipy.optimize` methods. SciPy contains a collection of scientific computing functions as an extension of Python. Chances are that these open source algorithms might fit into your applications off the shelf.

Root-finding scalar functions

Some root-finding functions that can be found in the `scipy.optimize` modules include `bisect`, `newton`, `brentq`, and `ridder`. Let's set up the examples that we have discussed in the *Incremental search* section using the implementations by SciPy:

```
In [ ]:
    """
    Documentation at
    http://docs.scipy.org/doc/scipy/reference/optimize.html
    """
    import scipy.optimize as optimize

    y = lambda x: x**3 + 2.*x**2 - 5.
    dy = lambda x: 3.*x**2 + 4.*x

    # Call method: bisect(f, a, b[, args, xtol, rtol, maxiter, ...])
    print("Bisection method:", optimize.bisect(y, -5., 5., xtol=0.00001))

    # Call method: newton(func, x0[, fprime, args, tol, ...])
    print("Newton's method:", optimize.newton(y, 5., fprime=dy))
    # When fprime=None, then the secant method is used.
    print("Secant method:", optimize.newton(y, 5.))

    # Call method: brentq(f, a, b[, args, xtol, rtol, maxiter, ...])
    print("Brent's method:", optimize.brentq(y, -5., 5.))
```

When we run the preceding code, the following output is generated:

```
Out[ ]:
    Bisection method: 1.241903305053711
    Newton's method: 1.2418965630344798
    Secant method: 1.2418965630344803
    Brent's method: 1.241896563034559
```

We can see that the SciPy implementation gives us very similar answers to our derived ones.

It should be noted that SciPy has a set of well-defined conditions for every implementation. For example, the function call of the bisection routine in the documentation is given as follows:

```
scipy.optimize.bisect(f, a, b, args=(), xtol=1e-12,
rtol=4.4408920985006262e-16, maxiter=100, full_output=False, disp=True)
```

The function will strictly evaluate the function, *f*, to return a zero of the function. *f(a)* and *f(b)* cannot have the same signs. In certain scenarios, it is difficult to fulfill these constraints. For example, in solving for nonlinear implied volatility models, volatility values cannot be negative. In active markets, finding a root or a zero of the volatility function is almost impossible without modifying the underlying implementation. In such cases, implementing our own root-finding methods might perhaps give us more authority over how our application should behave.

General nonlinear solvers

The `scipy.optimize` module also contains multidimensional general solvers that we can use to our advantage. The `root` and `fsolve` functions are some examples with the following function properties:

- `root(fun, x0[, args, method, jac, tol, ...])`: This finds a root of a vector function.
- `fsolve(func, x0[, args, fprime, ...])`: This finds the roots of a function.

The outputs are returned as dictionary objects. Using our example as input to these functions, we will get the following output:

```
In [ ]:
    import scipy.optimize as optimize

    y = lambda x: x**3 + 2.*x**2 - 5.
    dy = lambda x: 3.*x**2 + 4.*x

    print(optimize.fsolve(y, 5., fprime=dy))
Out[ ]:
    [1.24189656]
In [ ]:
    print(optimize.root(y, 5.))
Out[ ]:
    fjac: array([[-1.]])
     fun: array([3.55271368e-15])
 message: 'The solution converged.'
    nfev: 12
     qtf: array([-3.73605502e-09])
       r: array([-9.59451815])
  status: 1
 success: True
       x: array([1.24189656])
```

Using an initial guess value of 5, our solution converged to the root at 1.24189656, which is pretty close to the answers we've had so far. What happens when we choose a value on the other side of the graph? Let's use an initial guess value of −5:

```
In [ ]:
    print(optimize.fsolve(y, -5., fprime=dy))
Out[ ]:
    [-1.33306553]
    c:\python37\lib\site-packages\scipy\optimize\minpack.py:163:
RuntimeWarning: The iteration is not making good progress, as measured by
the
    improvement from the last ten iterations.
    warnings.warn(msg, RuntimeWarning)
In [ ]:
    print(optimize.root(y, -5.))
Out[ ]:
    fjac: array([[-1.]])
     fun: array([-3.81481496])
 message: 'The iteration is not making good progress, as measured by the \n
improvement from the last ten iterations.'
    nfev: 28
     qtf: array([3.81481521])
       r: array([-0.00461503])
  status: 5
 success: False
       x: array([-1.33306551])
```

As you can see from the display output, the algorithms did not converge and returned a root that is a little bit different from our previous answers. If we take a look at the equation on a graph, there are a number of points along the curve that lie very close to the root. A root-finder would be needed to obtain the desired level of accuracy, while solvers attempt to solve for the nearest answer in the fastest time.

Summary

In this chapter, we briefly discussed the persistence of nonlinearity in economics and finance. We looked at some nonlinear models that are commonly used in finance to explain certain aspects of data left unexplained by linear models: the Black-Scholes implied volatility model, Markov switching model, threshold model, and smooth transition models.

In Black-Scholes implied-volatility modeling, we discussed the volatility smile, which was made up of implied volatilities derived via the Black-Scholes model from the market prices of call or put options for a particular maturity. You may be interested enough to seek the lowest implied-volatility value possible, which can be useful for inferring theoretical prices and comparing against market prices for potential opportunities. However, since the curve is nonlinear, linear algebra cannot adequately solve for the optimal point. To do so, we will require the use of root-finding methods.

Root-finding methods attempt to find the root of a function or its zero. We discussed common root-finding methods, such as the bisection method, Newton's method, and the secant method. Using a combination of root-finding algorithms may help us to find roots of complex functions faster. One such example is Brent's method.

We explored functionalities in the `scipy.optimize` module that contains these root-finding methods, albeit with constraints. One of these constraints requires that the two boundary input values be evaluated with a pair of a negative value and a positive value for the solution to converge successfully. In implied volatility modeling, this evaluation is almost impossible, since volatilities do not have negative values. Implementing our own root-finding methods might perhaps give us more authority over how our application should perform.

Using general solvers is another way to find roots. They may also converge to our solution more quickly, but such convergence is not guaranteed by the initial given values.

Nonlinear modeling and optimization are inherently a complex task, and there is no universal solution or sure way to reach a conclusion. This chapter serves to introduce nonlinearity studies for finance in general.

In the next chapter, we will take a look at numerical methods commonly used for options pricing. By pairing a numerical procedure with a root-finding algorithm, we will learn how to build an implied volatility model with the market prices of an equity option.

4
Numerical Methods for Pricing Options

A derivative is a contract whose payoff depends on the value of some underlying asset. In cases where closed-form derivative pricing may be complex or even impossible, numerical procedures excel. A numerical procedure is the use of iterative computational methods in attempting to converge to a solution. One such basic implementation is a binomial tree. In a binomial tree, a node represents the state of an asset at a certain point of time associated with a price. Each node leads to two other nodes in the next time step. Similarly, in a trinomial tree, each node leads to three other nodes in the next time step. However, as the number of nodes or the time steps of trees increase, so do the computational resources that are consumed. Lattice pricing attempts to solve this problem by storing only the new information at each time step, while reusing values where possible.

In finite difference pricing, the nodes of the tree can also be represented as a grid. The terminal values on the grid consist of terminal conditions, while the edges of the grid represent boundary conditions in asset pricing. We will discuss the explicit method, implicit method, and the Crank-Nicolson method of the finite difference schemes to determine the price of an asset.

Although vanilla options and certain exotics such as European barrier options and lookback options can be found to have a closed-form solution, other exotic products such as Asian options do not contain a closed-form solution. In these cases, the pricing of options can be used with numerical procedures.

In this chapter, we will cover the following topics:

- Pricing European and American options using a binomial tree
- Using a Cox-Ross-Rubinstein binomial tree
- Pricing options using a Leisen-Reimer tree
- Pricing options using a trinomial tree

- Deriving Greeks from a tree for free
- Pricing options using a binomial and trinomial lattice
- Finite differences with the explicit, implicit, and Crank-Nicolson method
- Implied volatility modeling using an LR tree and the bisection method

Introduction to options

An **option** is a derivative of an asset that gives an owner the right but not the obligation to transact the underlying asset at a certain date for a certain price, known as the maturity date and strike price, respectively.

A **call option** gives the buyer the right to buy an asset by a certain date for a certain price. A seller or writer of a call option is obligated to sell the underlying security to the buyer at the agreed price, should the buyer exercise his/her rights on the agreed date.

A **put option** gives the buyer the right to sell the underlying asset by a certain date for a certain price. A seller or writer of a put option is obligated to buy the underlying security from the buyer at the agreed price, should the buyer exercise his/her rights on the agreed date.

The most common options that are available are the European options and American options. Other exotic options include Bermudan options and Asian options. This chapter will deal mainly with European and American options. A European option can only be exercised on the maturity date. An American option, on the other hand, may be exercised at any time throughout the lifetime of the option.

Binomial trees in option pricing

In the binomial option pricing model, the underlying security at one time period, represented as a node with a given price, is assumed to traverse to two other nodes in the next time step, representing an up state and a down state. Since options are derivatives of the underlying asset, the binomial pricing model tracks the underlying conditions on a discrete-time basis. Binomial option pricing can be used to value European options, American options, as well as Bermudan options.

The initial value of the root node is the spot price S_0 of the underlying security with a risk-neutral probability of increase q, and a risk-neutral probability of loss $1-q$, at the next time step. Based on these probabilities, the expected values of the security are calculated for each state of price increase or decrease for every time step. The terminal nodes represent every value of the expected security prices for every combination of up states and down states. We can then calculate the value of the option at every node, traverse the tree by risk-neutral expectations, and after discounting from the forward interest rates, we can derive the value of the call or put option.

Pricing European options

Consider a two-step binomial tree. A non-dividend paying stock price starts at $50, and, in each of the two time steps, the stock may go up by 20 percent or go down by 20 percent. Suppose that the risk-free rate is five percent per annum and that the time to maturity, T, is two years. We would like to find the value of a European put option with a strike price K of $52. The following diagram shows the pricing of the stock and the payoffs at the terminal nodes using a binomial tree:

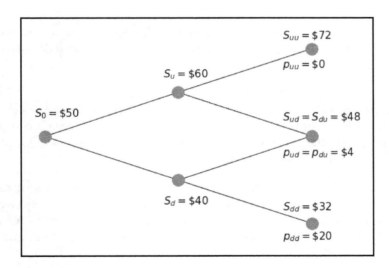

Here, the nodes are calculated as follows:

$$\text{Spot Price, } S_0 = 50$$

$$\text{Probability of price at up state, } u = 1.2$$

Probability of price at state, $d = 0.8$

$$S_u = 50(1.2) = 60$$

$$S_d = 50(0.8) = 40$$

$$S_{uu} = 50(1.2)^2 = 72$$

$$S_{ud} = S_{du} = 50(1.2)(0.8) = 48$$

$$S_{dd} = 50(0.8)^2 = 32$$

At the terminal nodes, the payoff from exercising a European call option is given as follows:

$$c_t = max(0, S_t - K)$$

In the case of a European put option, the payoff is as follows:

$$p_t = max(0, K - S_t)$$

 European call and put options are usually denoted by lowercase letters, c and p, while American call and put options are usually denoted by uppercase letters, C and P.

From the option payoff values, we can then traverse the binomial tree backward to the current time, and after discounting from the risk-free rate we will obtain our present value of the option. Traversing the tree backward takes into account the risk-neutral probabilities of the option's up states and down states.

We may assume that investors are indifferent to risk and that expected returns on all assets are equal. In the case of investing in stocks by risk-neutral probability, the payoff from holding the stock and taking into account the up and down state possibilities would be equal to the continuously compounded risk-free rate expected in the next time step, as follows:

$$e^{rt} = qu + (1 - q)d$$

The risk-neutral probability q of investing in the stock can be rewritten as follows:

$$q = \frac{e^{rt} - d}{u - d}$$

Are these formulas relevant to stocks? What about futures?
Unlike investing in stocks, investors do not have to make an upfront payment to take a position in a futures contract. In a risk-neutral sense, the expected growth rate from holding a futures contract is zero, and the risk-neutral probability q of investing in futures can be rewritten as follows:

$$q = \frac{1 - d}{u - d}$$

Let's calculate the risk-neutral probability q of the stock given in the preceding example:

```
In [ ]:
    import math

    r = 0.05
    T = 2
    t = T/2
    u = 1.2
    d = 0.8

    q = (math.exp(r*t)-d)/(u-d)
In [ ]:
    print('q is', q)
Out[ ]:
    q is 0.6281777409400603
```

The payoffs of exercising the European put option at the terminal nodes are $0, $4, and $20 at the respective states. The present value of the put option can be priced as follows:

$$p_t = e^{-rT} \left[0(q)^2 + 2(48)(q)(1 - q) + 20(1 - q)^2 \right]$$

This gives us the put option price as $4.19. The two-step binomial tree to value a European put option with payoffs at each node is illustrated in the following graph:

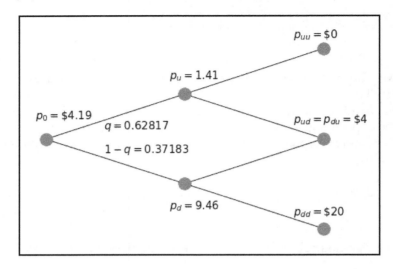

Writing the StockOption base class

Before going any further and implementing the various pricing models that we are about to discuss, let's create a `StockOption` class to store and calculate the common attributes of the stock option that will be reused throughout this chapter:

```
In [ ]:
    import math

    """
    Stores common attributes of a stock option
    """
    class StockOption(object):
        def __init__(
            self, S0, K, r=0.05, T=1, N=2, pu=0, pd=0,
            div=0, sigma=0, is_put=False, is_am=False):
            """
            Initialize the stock option base class.
            Defaults to European call unless specified.
            :param S0: initial stock price
            :param K: strike price
            :param r: risk-free interest rate
            :param T: time to maturity
            :param N: number of time steps
```

```
:param pu: probability at up state
:param pd: probability at down state
:param div: Dividend yield
:param is_put: True for a put option,
        False for a call option
:param is_am: True for an American option,
        False for a European option
"""
self.S0 = S0
self.K = K
self.r = r
self.T = T
self.N = max(1, N)
self.STs = [] # Declare the stock prices tree

""" Optional parameters used by derived classes """
self.pu, self.pd = pu, pd
self.div = div
self.sigma = sigma
self.is_call = not is_put
self.is_european = not is_am

@property
def dt(self):
    """ Single time step, in years """
    return self.T/float(self.N)

@property
def df(self):
    """ The discount factor """
    return math.exp(-(self.r-self.div)*self.dt)
```

The current underlying price, strike price, risk-free rate, time to maturity, and number of time steps are compulsory common attributes for pricing options. The delta of the time step dt and the discount factor df are computed as properties of the class and may be overwritten by implementing classes if needed.

A class for European options using a binomial tree

The Python implementation of the binomial option pricing model of a European option is given as the `BinomialEuropeanOption` class, inheriting the common attributes of the option from the `StockOption` class. The implementations of the methods in this class are as follows:

1. The `price()` method of the `BinomialEuropeanOption` class is the entry point for all the instances of this class

2. It invokes the `setup_parameters()` method to set up the required model parameters, and then calls the `init_stock_price_tree()` method to simulate the expected values of the stock prices for the period up until T

3. Finally, the `begin_tree_traversal()` method is called to initialize the payoff array and store the discounted payoff values, as it traverses the binomial tree back to the present time

4. The payoff tree nodes are returned as a NumPy array object, where the present value of the European option is found at the initial node

The class implementation of `BinomialEuropeanOption` is given in the following Python code:

```
In [ ]:
    import math
    import numpy as np
    from decimal import Decimal

    """
    Price a European option by the binomial tree model
    """
    class BinomialEuropeanOption(StockOption):

        def setup_parameters(self):
            # Required calculations for the model
            self.M = self.N+1   # Number of terminal nodes of tree
            self.u = 1+self.pu  # Expected value in the up state
            self.d = 1-self.pd  # Expected value in the down state
            self.qu = (math.exp(
                (self.r-self.div)*self.dt)-self.d)/(self.u-self.d)
            self.qd = 1-self.qu

        def init_stock_price_tree(self):
            # Initialize terminal price nodes to zeros
            self.STs = np.zeros(self.M)
```

```
            # Calculate expected stock prices for each node
            for i in range(self.M):
                self.STs[i] = self.S0 * \
                    (self.u**(self.N-i)) * (self.d**i)

        def init_payoffs_tree(self):
            """
            Returns the payoffs when the option
            expires at terminal nodes
            """
            if self.is_call:
                return np.maximum(0, self.STs-self.K)
            else:
                return np.maximum(0, self.K-self.STs)

        def traverse_tree(self, payoffs):
            """
            Starting from the time the option expires, traverse
            backwards and calculate discounted payoffs at each node
            """
            for i in range(self.N):
                payoffs = (payoffs[:-1]*self.qu +
                            payoffs[1:]*self.qd)*self.df

            return payoffs

        def begin_tree_traversal(self):
            payoffs = self.init_payoffs_tree()
            return self.traverse_tree(payoffs)

        def price(self):
            """ Entry point of the pricing implementation """
            self.setup_parameters()
            self.init_stock_price_tree()
            payoffs = self.begin_tree_traversal()
            # Option value converges to first node
            return payoffs[0]
```

Let's take the values from the two-step binomial tree example we discussed earlier to price the European put option:

```
In [ ]:
    eu_option = BinomialEuropeanOption(
        50, 52, r=0.05, T=2, N=2, pu=0.2, pd=0.2, is_put=True)
In [ ]:
    print('European put option price is:', eu_option.price())
Out[ ]:
    European put option price is: 4.1926542806038585
```

Using the binomial option pricing model gives us a present value of $4.19 for the European put option.

A class for American options using a binomial tree

Unlike European options, which can only be exercised at maturity, American options can be exercised at any time during their lifetime.

To implement the pricing of American options in Python, in the same way we did with the `BinomialEuropeanOption` class, create a class named `BinomialTreeOption` that inherits the `StockOption` class. The parameters that are used in the `setup_parameters()` method remain the same, except for the removal of an unused M parameter.

The methods that are used in American options are as follows:

- `init_stock_price_tree`: Uses a two-dimensional NumPy array to store the expected returns of the stock prices for all time steps. This information is used to calculate the payoff values from exercising the option at each period. This method is written as follows:

```
def init_stock_price_tree(self):
    # Initialize a 2D tree at T=0
    self.STs = [np.array([self.S0])]

    # Simulate the possible stock prices path
    for i in range(self.N):
        prev_branches = self.STs[-1]
        st = np.concatenate(
            (prev_branches*self.u,
             [prev_branches[-1]*self.d]))
        self.STs.append(st) # Add nodes at each time step
```

- `init_payoffs_tree`: Creates the payoff tree as a two-dimensional NumPy array, starting with the intrinsic values of the option at maturity. This method is written as follows:

```
def init_payoffs_tree(self):
    if self.is_call:
        return np.maximum(0, self.STs[self.N]-self.K)
    else:
        return np.maximum(0, self.K-self.STs[self.N])
```

- `check_early_exercise`: Returns the maximum payoff values between exercising the American option early and not exercising the option at all. This method is written as follows:

```
def check_early_exercise(self, payoffs, node):
    if self.is_call:
        return np.maximum(payoffs, self.STs[node] - self.K)
    else:
        return np.maximum(payoffs, self.K - self.STs[node])
```

- `traverse_tree`: This also includes the invocation of the `check_early_exercise()` method to check whether it is optimal to exercise an American option early at every time step. This method is written as follows:

```
def traverse_tree(self, payoffs):
    for i in reversed(range(self.N)):
        # The payoffs from NOT exercising the option
        payoffs = (payoffs[:-1]*self.qu +
                    payoffs[1:]*self.qd)*self.df

        # Payoffs from exercising, for American options
        if not self.is_european:
            payoffs = self.check_early_exercise(payoffs,i)

    return payoffs
```

The implementation of the `begin_tree_traversal()` and the `price()` methods remains the same.

The `BinomialTreeOption` class can price both European and American options when the `is_put` keyword argument is set to `False` or `True` during instantiation of the class, respectively.

The following code is for pricing the American option:

```
In [ ]:
    am_option = BinomialTreeOption(50, 52,
        r=0.05, T=2, N=2, pu=0.2, pd=0.2, is_put=True, is_am=True)
In [ ]:
    print('American put option price is:', am_option.price())
Out[ ]:
    American put option price is: 5.089632474198373
```

The American put option is priced at $5.0896. Since American options can be exercised at any time and European options can only be exercised at maturity, this added flexibility of American options increases their value over European options in certain circumstances.

For American call options on an underlying asset that does not pay dividends, there might not be an extra value over its European call option counterpart. Because of the time value of money, it costs more to exercise the American call option today before the expiration at the strike price than at a future time with the same strike price. For an in-the-money American call option, exercising the option early loses the benefit of protection against adverse price movement below the strike price, as well as its intrinsic time value. With no entitlement of dividend payments, there are no incentives to exercise American call options early.

The Cox–Ross–Rubinstein model

In the preceding examples, we assumed that the underlying stock price would increase by 20 percent and decrease by 20 percent in the respective *u* up state and *d* down state. The **Cox-Ross-Rubinstein** (**CRR**) model proposes that, over a short period of time in the risk-neutral world, the binomial model matches the mean and variance of the underlying stock. The volatility of the underlying stock, or the standard deviation of returns of the stock, is taken into account as follows:

$$u = e^{\sigma\sqrt{\Delta t}}$$

$$d = \frac{1}{u} = e^{-\sigma\sqrt{\Delta t}}$$

A class for the CRR binomial tree option pricing model

The implementation of the binomial CRR model remains the same as the binomial tree we discussed earlier, with the exception of the u and d model parameters.

In Python, let's create a class named `BinomialCRROption` and simply inherit the `BinomialTreeOption` class. Then, all that we need to do is override the `setup_parameters()` method with values from the CRR model.

Instances of the `BinomialCRROption` object will invoke the `price()` method, which invokes all other methods of the parent `BinomialTreeOption` class, except the overwritten `setup_parameters()` method:

```
In [ ]:
    import math

    """
    Price an option by the binomial CRR model
```

```
"""
class BinomialCRROption(BinomialTreeOption):
    def setup_parameters(self):
        self.u = math.exp(self.sigma * math.sqrt(self.dt))
        self.d = 1./self.u
        self.qu = (math.exp((self.r-self.div)*self.dt) -
                    self.d)/(self.u-self.d)
        self.qd = 1-self.qu
```

Again, consider the two-step binomial tree. The non-dividend paying stock has a current price of $50 and a volatility of 30 percent. Suppose that the risk-free rate is five percent per annum and the time to maturity T is two years. We would like to find the value of a European put option with a strike price K of $52 by the CRR model:

```
In [ ]:
    eu_option = BinomialCRROption(
        50, 52, r=0.05, T=2, N=2, sigma=0.3, is_put=True)
In [ ]:
    print('European put:', eu_option.price())
Out[ ]:
    European put: 6.245708445206436
In [ ]:
    am_option = BinomialCRROption(50, 52,
        r=0.05, T=2, N=2, sigma=0.3, is_put=True, is_am=True)
In [ ]:
    print('American put option price is:', am_option.price())
Out[ ]:
    American put option price is: 7.428401902704834
```

By using the CRR two-step binomial tree model, the price of the European put option and the American put option are $6.2457 and $7.4284, respectively.

Using a Leisen-Reimer tree

In the binomial models we discussed earlier, we made several assumptions about the probability of up and down states, as well as the resulting risk-neutral probabilities. Besides the binomial model with CRR parameters that we discussed, other forms of parameterization that are discussed widely in mathematical finance include the Jarrow-Rudd parameterization, Tian parameterization, and Leisen-Reimer parameterization. Let's take a look at the Leisen-Reimer model in detail.

Dr. Dietmar Leisen and Matthias Reimer proposed a binomial tree model with the purpose of approximating to the Black-Scholes solution as the number of step increases. It is known as the **Leisen-Reimer** (**LR**) tree, and the nodes do not recombine at every alternate step. It uses an inversion formula to achieve better accuracy during tree traversal.

A detailed explanation of the formula is given in the paper *Binomial Models For Option Valuation - Examining And Improving Convergence*, March 1995, which is available at `http://papers.ssrn.com/sol3/papers.cfm?abstract_id=5976`. We will be using method two of the Peizer and Pratt inversion function f with the following characteristic parameters:

$$f(z, j(n)) = 0.5 \mp \left[0.25 - 0.25 exp\left\{ -\left(\frac{z}{n + \frac{1}{3} + \frac{0.1}{n+1}} \right)^2 \left(n + \frac{1}{6}\right) \right\} \right]^{1/2}$$

$$j(n) = \begin{cases} n, \text{if } n \text{ is even} \\ n + 1, \text{if } n \text{ is odd} \end{cases}$$

$$p' = f(d_1, j(n))$$

$$p = f(d_2, j(n))$$

$$d_1 = \frac{log\left(\frac{S_0}{K}\right) + \left((r - y) + \frac{\sigma^2}{2}\right)T}{\sigma\sqrt{T}}$$

$$d_2 = \frac{log\left(\frac{S_0}{K}\right) - \left((r - y) + \frac{\sigma^2}{2}\right)T}{\sigma\sqrt{T}}$$

$$u = e^{(r-y)\Delta t}\frac{p'}{p}$$

$$d = \frac{e^{(r-y)\Delta y} - pu}{1 - p}$$

The S_0 parameter is the current stock price, K is the strike price of the option, σ is the annualized volatility of the underlying stock, T is the time to maturity of the option, r is the annualized risk-free rate, y is the dividend yield, and Δt is the time interval between each tree step.

A class for the LR binomial tree option pricing model

The Python implementation of the LR tree is given in the following `BinomialLROption`
class. Similar to the `BinomialCRROption` class, we simply inherit the
`BinomialTreeOption` class and override the variables in the `setup_parameters` method
with those of the LR tree model:

```
In [ ]:
    import math

    """
    Price an option by the Leisen-Reimer tree
    """
    class BinomialLROption(BinomialTreeOption):

        def setup_parameters(self):
            odd_N = self.N if (self.N%2 == 0) else (self.N+1)
            d1 = (math.log(self.S0/self.K) +
                ((self.r-self.div) +
                (self.sigma**2)/2.)*self.T)/\
              (self.sigma*math.sqrt(self.T))
            d2 = (math.log(self.S0/self.K) +
                ((self.r-self.div) -
                (self.sigma**2)/2.)*self.T)/\
              (self.sigma * math.sqrt(self.T))

            pbar = self.pp_2_inversion(d1, odd_N)
            self.p = self.pp_2_inversion(d2, odd_N)
            self.u = 1/self.df * pbar/self.p
            self.d = (1/self.df-self.p*self.u)/(1-self.p)
            self.qu = self.p
            self.qd = 1-self.p

        def pp_2_inversion(self, z, n):
            return .5 + math.copysign(1, z)*\
                math.sqrt(.25 - .25*
                    math.exp(
                        -((z/(n+1./3.+.1/(n+1)))**2.)*(n+1./6.)
                    )
                )
```

Using the same examples that we used previously, we can price the options using an LR
tree:

```
In [ ]:
    eu_option = BinomialLROption(
        50, 52, r=0.05, T=2, N=4, sigma=0.3, is_put=True)
```

```
In [ ]:
    print('European put:', eu_option.price())
Out[ ]:
    European put: 5.878650106601964
In [ ]:
    am_option = BinomialLROption(50, 52,
        r=0.05, T=2, N=4, sigma=0.3, is_put=True, is_am=True)
In [ ]:
    print('American put:', am_option.price())
Out[ ]:
    American put: 6.763641952939979
```

By using the LR binomial tree model with four time steps, the price of the European put option and the American put option are \$5.87865 and \$6.7636, respectively.

The Greeks for free

In the binomial tree pricing models that we have covered so far, we traversed up and down the tree at each point in time to determine the node values. From the information at each node, we can reuse these computed values easily. One such use is the computation of Greeks.

The Greeks measure the sensitivities of the price of derivatives, such as options with respect to changes in the parameters of the underlying asset, often represented by Greek letters. In mathematical finance, the common names associated with Greeks include alpha, beta, delta, gamma, vega, theta, and rho.

Two particularly useful Greeks for options are delta and gamma. Delta measures the sensitivity of the option price with respect to the underlying asset price. Gamma measures the rate of change in delta with respect to the underlying price.

As shown in the following diagram, an additional layer of nodes is added around our original two-step tree to make it a four-step tree, which extends two steps backward in time. Even with additional terminal payoff nodes, all nodes will contain the same information as our original two-step tree. Our option value of interest is now located in the middle of the tree at **t=0**:

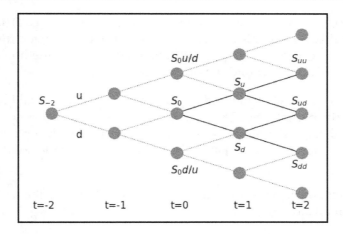

Notice that at **t=0** there exists two additional nodes' worth of information that we can use to compute the delta formula, as follows:

$$\text{delta} = \frac{v_{up} - v_{down}}{S_0 u/d - S_0 d/u}$$

The delta formula states that the difference in the option prices in the up and down state is represented as a unit of the difference between the respective stock prices at time **t=0**.

Conversely, the gamma formula can be computed as follows:

$$\gamma = \frac{\dfrac{v_{up} - v_0}{S_{0,up} - S_0} - \dfrac{v_0 - v_{down}}{S_0 - S_{0,down}}}{\dfrac{S_0 + S_{0,up}}{2} - \dfrac{S_0 + S_{0,down}}{2}}$$

The gamma formula states that the difference of deltas between the option prices in the up node and the down node against the initial node value are computed as a unit of the differences in price of the stock at the respective states.

A class for Greeks with the LR binomial tree

To illustrate the computation of Greeks with the LR tree, let's create a new class named `BinomialLRWithGreeks` that inherits the `BinomialLROption` class with our own implementation of the `price` method.

In the `price` method, we will start by calling the `setup_parameters()` method of the parent class to initialize all variables required by the LR tree. However, this time, we will also call the `new_stock_price_tree()` method, which is a new method that's used to create an extra layer of nodes around the original tree.

The `begin_tree_traversal()` method is called to perform the usual LR tree implementation in the parent class. The returned NumPy array object now contains information on the three nodes at **t=0**, where the middle node is the option price. The payoffs in the up and down states at **t=0** are in the first and last index of the array, respectively.

With this information, the `price()` method computes and returns the option price, the delta, and the gamma values together:

```python
In [ ]:
    import numpy as np

    """
    Compute option price, delta and gamma by the LR tree
    """
    class BinomialLRWithGreeks(BinomialLROption):

        def new_stock_price_tree(self):
            """
            Creates an additional layer of nodes to our
            original stock price tree
            """
            self.STs = [np.array([self.S0*self.u/self.d,
                                  self.S0,
                                  self.S0*self.d/self.u])]

            for i in range(self.N):
                prev_branches = self.STs[-1]
                st = np.concatenate((prev_branches*self.u,
                                     [prev_branches[-1]*self.d]))
                self.STs.append(st)

        def price(self):
            self.setup_parameters()
            self.new_stock_price_tree()
            payoffs = self.begin_tree_traversal()

            # Option value is now in the middle node at t=0
            option_value = payoffs[len(payoffs)//2]

            payoff_up = payoffs[0]
```

```
        payoff_down = payoffs[-1]
        S_up = self.STs[0][0]
        S_down = self.STs[0][-1]
        dS_up = S_up - self.S0
        dS_down = self.S0 - S_down

        # Calculate delta value
        dS = S_up - S_down
        dV = payoff_up - payoff_down
        delta = dV/dS

        # calculate gamma value
        gamma = ((payoff_up-option_value)/dS_up -
                (option_value-payoff_down)/dS_down) / \
            ((self.S0+S_up)/2. - (self.S0+S_down)/2.)

        return option_value, delta, gamma
```

Using the same example from the LR tree, we can compute the option values and Greeks for a European call and put option with 300 time steps:

```
In [ ]:
    eu_call = BinomialLRWithGreeks(50, 52, r=0.05, T=2, N=300, sigma=0.3)
    results = eu_call.price()
In [ ]:
    print('European call values')
    print('Price: %s\nDelta: %s\nGamma: %s' % results)
Out[ ]:
    European call values
    Price: 9.69546807138366
    Delta: 0.6392477816643529
    Gamma: 0.01764795890533088

In [ ]:
    eu_put = BinomialLRWithGreeks(
        50, 52, r=0.05, T=2, N=300, sigma=0.3, is_put=True)
    results = eu_put.price()
In [ ]:
    print('European put values')
    print('Price: %s\nDelta: %s\nGamma: %s' % results)
Out[ ]:
    European put values
    Price: 6.747013809252746
    Delta: -0.3607522183356649
    Gamma: 0.017647958905312
```

As we can see from the `price()` method and results, we managed to obtain additional information on Greeks from the modified binomial tree without any extra overhead in computational complexity.

Trinomial trees in option pricing

In the binomial tree, each node leads to two other nodes in the next time step. Similarly, in a trinomial tree, each node leads to three other nodes in the next time step. Besides having up and down states, the middle node of the trinomial tree indicates no change in state. When extended over more than two time steps, the trinomial tree can be thought of as a recombining tree, where the middle nodes always retain the same values as the previous time step.

Let's consider the Boyle trinomial tree, where the tree is calibrated so that the probability of up, down, and flat movements, u, d, and m with risk-neutral probabilities q_u, q_d, and q_m are as follows:

$$u = e^{\sigma\sqrt{2\Delta t}}$$

$$d = \frac{1}{u} = e^{-\sigma\sqrt{2\Delta t}}$$

$$m = ud = 1$$

$$q_u = \left(\frac{e^{(r-v)\frac{\Delta t}{2}} - e^{\sigma\sqrt{\frac{\Delta t}{2}}}}{e^{\sigma\sqrt{\frac{\Delta t}{2}}} - e^{-\sigma\sqrt{\frac{\Delta t}{2}}}} \right)^2$$

$$q_d = \left(\frac{e^{\sigma\sqrt{\frac{\Delta t}{2}}} - e^{(r-v)\frac{\Delta t}{2}}}{e^{\sigma\sqrt{\frac{\Delta t}{2}}} - e^{-\sigma\sqrt{\frac{\Delta t}{2}}}} \right)^2$$

$$q_m = 1 - q_u - q_d$$

We can see that $ud = e^{\sigma\sqrt{2\Delta t}}e^{-\sigma\sqrt{2\Delta t}}$ recombines to $m = 1$. With calibration, the no state movement m grows at a flat rate of 1 instead of at the risk-free rate. The variable v is the annualized dividend yield, and σ is the annualized volatility of the underlying stock.

In general, with an increased number of nodes to process, a trinomial tree gives better accuracy than the binomial tree when fewer time steps are modeled, saving on computation speed and resources. The following diagram illustrates the stock price movements of a trinomial tree with two time steps:

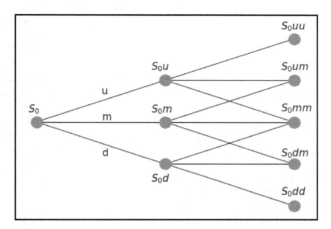

A class for the trinomial tree option pricing model

Let's create a `TrinomialTreeOption` class, inheriting from the `BinomialTreeOption` class.

The methods for the `TrinomialTreeOption` are provided as follows:

- The `setup_parameters()` method implements the model parameters of the trinomial tree. This method is written as follows:

```
def setup_parameters(self):
    """ Required calculations for the model """
    self.u = math.exp(self.sigma*math.sqrt(2.*self.dt))
    self.d = 1/self.u
    self.m = 1
    self.qu = ((math.exp((self.r-self.div) *
                    self.dt/2.) -
              math.exp(-self.sigma *
                    math.sqrt(self.dt/2.))) /
              (math.exp(self.sigma *
                    math.sqrt(self.dt/2.)) -
              math.exp(-self.sigma *
                    math.sqrt(self.dt/2.))))**2
    self.qd = ((math.exp(self.sigma *
```

```
                              math.sqrt(self.dt/2.)) -
                  math.exp((self.r-self.div) *
                          self.dt/2.)) /
                  (math.exp(self.sigma *
                          math.sqrt(self.dt/2.)) -
                  math.exp(-self.sigma *
                          math.sqrt(self.dt/2.)))) **2.

         self.qm = 1 - self.qu - self.qd
```

- The `init_stock_price_tree()` method sets up the trinomial tree to include the flat movement of stock prices. This method is written as follows:

```
def init_stock_price_tree(self):
    # Initialize a 2D tree at t=0
    self.STs = [np.array([self.S0])]

    for i in range(self.N):
        prev_nodes = self.STs[-1]
        self.ST = np.concatenate(
            (prev_nodes*self.u, [prev_nodes[-1]*self.m,
                                 prev_nodes[-1]*self.d]))
        self.STs.append(self.ST)
```

- The `traverse_tree()` method takes into account the middle node after discounting the payoff:

```
def traverse_tree(self, payoffs):
    # Traverse the tree backwards
    for i in reversed(range(self.N)):
        payoffs = (payoffs[:-2] * self.qu +
                   payoffs[1:-1] * self.qm +
                   payoffs[2:] * self.qd) * self.df

        if not self.is_european:
            payoffs = self.check_early_exercise(payoffs,i)

    return payoffs
```

- Using the same example of the binomial tree, we get the following result:

```
In [ ]:
   eu_put = TrinomialTreeOption(
        50, 52, r=0.05, T=2, N=2, sigma=0.3, is_put=True)
In [ ]:
   print('European put:', eu_put.price())
Out[ ]:
```

```
    European put: 6.573565269142496
In [ ]:
    am_option = TrinomialTreeOption(50, 52,
        r=0.05, T=2, N=2, sigma=0.3, is_put=True, is_am=True)
In [ ]:
    print('American put:', am_option.price())
Out[ ]:
    American put: 7.161349217272585
```

By the trinomial tree model, we obtain prices of $6.57 and $7.16 for the European and American put options, respectively.

Lattices in option pricing

In binomial trees, each node recombines at every alternative node. In trinomial trees, each node recombines at every other node. This property of recombining trees can also be represented as lattices to save memory without recomputing and storing recombined nodes.

Using a binomial lattice

We will create a binomial lattice from the binomial CRR tree since at every alternate up and down nodes, the prices recombine to the same probability of $ud=1$. In the following diagram, $\mathbf{S_u}$ and $\mathbf{S_d}$ recombine with $\mathbf{S_{du}} = \mathbf{S_{ud}} = \mathbf{S_0}$. The tree can now be represented as a single list:

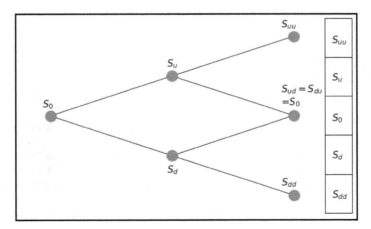

For a *N*-step binomial tree, a list of size *2N +1* is required to contain the information on the underlying stock prices. For European option pricing, the odd nodes of payoffs from the list represent the option value upon maturity. The tree traverses backward to obtain the option value. For American option pricing, as the tree traverses backward, both ends of the list shrink, and the odd nodes represent the associated stock prices for any time step. Payoffs from the earlier exercise can then be taken into account.

A class for the CRR binomial lattice option pricing model

Let's convert the binomial tree pricing into a lattice by CRR. We can inherit from the BinomialCRROption class (which in turn inherits the BinomialTreeOption class) and create a new class named BinomialCRRLattice, as follows:

```
In [ ]:
    import numpy as np

    class BinomialCRRLattice(BinomialCRROption):

        def setup_parameters(self):
            super(BinomialCRRLattice, self).setup_parameters()
            self.M = 2*self.N + 1

        def init_stock_price_tree(self):
            self.STs = np.zeros(self.M)
            self.STs[0] = self.S0 * self.u**self.N

            for i in range(self.M)[1:]:
                self.STs[i] = self.STs[i-1]*self.d

        def init_payoffs_tree(self):
            odd_nodes = self.STs[::2]   # Take odd nodes only
            if self.is_call:
                return np.maximum(0, odd_nodes-self.K)
            else:
                return np.maximum(0, self.K-odd_nodes)

        def check_early_exercise(self, payoffs, node):
            self.STs = self.STs[1:-1]   # Shorten ends of the list
            odd_STs = self.STs[::2]   # Take odd nodes only
            if self.is_call:
                return np.maximum(payoffs, odd_STs-self.K)
            else:
```

```
                    return np.maximum(payoffs, self.K-odd_STs)
```

The following methods are overwritten with the implementation of the lattice while retaining the behavior of all the other pricing functions:

- `setup_parameters`: Overrides the parent method to initialize the CRR parameters of the parent class, as well as declaring the new variable `M` as the list size

- `init_stock_price_tree`: Overrides the parent method to set up a one-dimensional NumPy array as the lattice with the `M` size

- `init_payoffs_tree` and `check_early_exercise`: Overrides the parent methods to take into account the payoffs at odd nodes only

Using the same stock information from our binomial CRR model example, we can price a European and American put option using the binomial lattice pricing:

```
In [ ]:
    eu_option = BinomialCRRLattice(
        50, 52, r=0.05, T=2, N=2, sigma=0.3, is_put=True)
In [ ] :
    print('European put:', eu_option.price())
Out[ ]:
    European put: 6.245708445206432
In [ ]:
    am_option = BinomialCRRLattice(50, 52,
        r=0.05, T=2, N=2, sigma=0.3, is_put=True, is_am=True)
In [ ] :
    print("American put:", am_option.price())
Out[ ]:
    American put: 7.428401902704828
```

By using the CRR binomial tree lattice pricing model, we obtain prices of \$6.2457 and \$7.428 for the European and American put options, respectively.

Using the trinomial lattice

The trinomial lattice works in very much the same way as the binomial lattice. Since each node recombines at every other node instead of alternate nodes, extracting odd nodes from the list is not necessary. Since the size of the list is the same as the one in the binomial lattice, there are no extra storage requirements in trinomial lattice pricing, as explained in the following diagram:

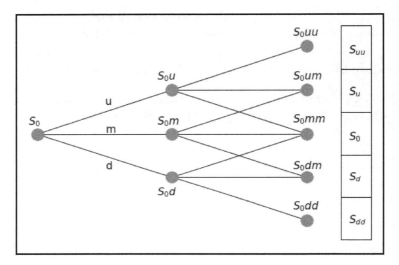

A class for the trinomial lattice option pricing model

In Python, let's create a class named `TrinomialLattice` for the trinomial lattice implementation that inherits from the `TrinomialTreeOption` class.

Just as we did for the `BinomialCRRLattice` class, the `setup_parameters`, `init_stock_price_tree`, `init_payoffs_tree`, and `check_early_exercise` methods are overwritten, without having to take into account the payoffs at odd nodes:

```
In [ ]:
    import numpy as np

    """
    Price an option by the trinomial lattice
    """
    class TrinomialLattice(TrinomialTreeOption):

        def setup_parameters(self):
```

```
        super(TrinomialLattice, self).setup_parameters()
        self.M = 2*self.N + 1

    def init_stock_price_tree(self):
        self.STs = np.zeros(self.M)
        self.STs[0] = self.S0 * self.u**self.N

        for i in range(self.M)[1:]:
            self.STs[i] = self.STs[i-1]*self.d

    def init_payoffs_tree(self):
        if self.is_call:
            return np.maximum(0, self.STs-self.K)
        else:
            return np.maximum(0, self.K-self.STs)

    def check_early_exercise(self, payoffs, node):
        self.STs = self.STs[1:-1]   # Shorten ends of the list
        if self.is_call:
            return np.maximum(payoffs, self.STs-self.K)
        else:
            return np.maximum(payoffs, self.K-self.STs)
```

Using the same examples as before, we can price the European and American options using the trinomial lattice model:

```
In [ ]:
    eu_option = TrinomialLattice(
        50, 52, r=0.05, T=2, N=2, sigma=0.3, is_put=True)
    print('European put:', eu_option.price())
Out[ ]:
    European put: 6.573565269142496
In [ ]:
    am_option = TrinomialLattice(50, 52,
        r=0.05, T=2, N=2, sigma=0.3, is_put=True, is_am=True)
    print('American put:', am_option.price())
Out[ ]:
    American put: 7.161349217272585
```

The output agrees with the results that were obtained from the trinomial tree option pricing model.

Finite differences in option pricing

Finite difference schemes are very much similar to trinomial tree option pricing, where each node is dependent on three other nodes with an up movement, a down movement, and a flat movement. The motivation behind the finite differencing is the application of the Black-Scholes **Partial Differential Equation** (PDE) framework (involving functions and their partial derivatives), where price $S(t)$ is a function of $f(S,t)$, with r as the risk-free rate, t as the time to maturity, and σ as the volatility of the underlying security:

$$rf = \frac{df}{dt} + rS\frac{df}{dS} + \frac{1}{2}\sigma^2 S^2 \frac{d^2 f}{dt^2}$$

The finite difference technique tends to converge faster than lattices and approximates complex exotic options very well.

To solve a PDE by finite differences working backward in time, a discrete-time grid of size M by N is set up to reflect asset prices over a course of time, so that S and t take on the following values at each point on the grid:

It follows that by grid notation, $f_{i,j}=f(\ idS,\ j\ dt)$. S_{max} is a suitably large asset price that cannot be reached by the maturity time, T. Thus dS and dt are intervals between each node in the grid, incremented by price and time, respectively. The terminal condition at expiration time T for every value of S is $max(S-K,\ 0)$ for a call option with strike K and $max(K-S,\ 0)$ for a put option. The grid traverses backward from the terminal conditions, complying with the PDE while adhering to the boundary conditions of the grid, such as the payoff from an earlier exercise.

The boundary conditions are defined values at the extreme ends of the nodes, where $i=0$ and $i=N$ for every time at t. Values at the boundaries are used to calculate the values of all other lattice nodes iteratively using the PDE.

A visual representation of the grid is given in the following diagram. As i and j increase from the top-left corner of the grid, the price S tends toward S_{max} (the maximum price possible) at the bottom-right corner of the grid:

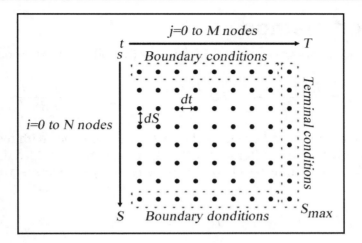

A number of ways to approximate the PDE are as follows:

- Forward difference:

$$\frac{df}{dS} = \frac{f_{i+1,j} - f_{i,j}}{dS}, \frac{df}{dt} = \frac{f_{i,j+1} - f_{i,j}}{dt}$$

- Backward difference:

$$\frac{df}{dS} = \frac{f_{i,j} - f_{i-1,j}}{dS}, \frac{df}{dt} = \frac{f_{i,j} - f_{i,j-1}}{dt}$$

- Central or symmetric difference:

$$\frac{df}{dS} = \frac{f_{i+1,j} - f_{i-1,j}}{2dS}, \frac{df}{dt} = \frac{f_{i,j+1} - f_{i,j-1}}{2dt}$$

- The second derivative:

$$\frac{d^2 f}{dS^2} = \frac{f_{i+1,j} - 2f_{i,j} + f_{i-1,j}}{dS^2}$$

Once we have the boundary conditions set up, we can now apply an iterative approach using the explicit, implicit, or Crank-Nicolson method.

The explicit method

The explicit method for approximating $f_{i,j}$ is given by the following equation:

$$rf_{i,j} = \frac{f_{i,j} - f_{i,j-1}}{dt} + ridS\frac{f_{i+1,j} - f_{i-1,j}}{2ds}\frac{1}{2}\sigma^2 j^2\frac{f_{i+1,j} + f_{i-1,j}}{dS^2}$$

Here, we can see that the first difference is the backward difference with respect to t, the second difference is the central difference with respect to S, and the third difference is the second-order difference with respect to S. When we rearrange the terms, we get the following equation:

$$f_{i,j} = a_i^* f_{i-1,j+1} + b_i^* f_{i,j+1} + c_i^* f_{i+1,j+1}$$

Where:

$$j = N-1, N-2, N = 3, \ldots, 2, 1, 0$$

$$i = 1, 2, 3, \ldots M-2, M-1$$

Then:

$$a_i^* = \frac{1}{2}dt(\sigma^2 i^2 - ri)$$

$$b_i^* = 1 - dt(\sigma^2 i^2 - ri)$$

$$c_i^* = \frac{1}{2}dt(\sigma^2 i^2 + ri)$$

The iterative approach of the explicit method can be visually represented by the following diagram:

Writing the finite difference base class

Since we will be writing the explicit, implicit, and Crank-Nicolson methods of finite differences in Python, let's write a base class that inherits the common properties and functions of all three methods.

We will create a class called `FiniteDifferences` that accepts and assigns all the required parameters in the `__init__` constructor method. The `price()` method is the entry point for invoking the specific finite difference scheme implementation, and will invoke these methods in the following order: `setup_boundary_conditions()`, `setup_coefficients()`, `traverse_grid()`, and `interpolate()`. These methods are explained as follows:

- `setup_boundary_conditions`: Sets up the boundary conditions of the grid structure as a NumPy two-dimensional array
- `setup_coefficients`: Sets up the necessary coefficients that are used for traversing the grid structure
- `traverse_grid`: Iterates the grid structure backward in time, storing the calculated values toward the first column of the grid
- `interpolate`: Using the final calculated values on the first column of the grid, this method will interpolate these values to find the option price that closely infers the initial stock price, `S0`

All of these methods are abstract methods that can be implemented by the derived classes. An exception type of `NotImplementedError` will be thrown should we forget to implement these methods.

The base class with the mandatory methods should look like this:

```
In [ ]:
    from abc import ABC, abstractmethod
    import numpy as np

    """
    Base class for sharing
    attributes and functions of FD
    """
    class FiniteDifferences(object):

        def __init__(
            self, S0, K, r=0.05, T=1,
            sigma=0, Smax=1, M=1, N=1, is_put=False
        ):
            self.S0 = S0
```

```
                self.K = K
                self.r = r
                self.T = T
                self.sigma = sigma
                self.Smax = Smax
                self.M, self.N = M, N
                self.is_call = not is_put
                self.i_values = np.arange(self.M)
                self.j_values = np.arange(self.N)
                self.grid = np.zeros(shape=(self.M+1, self.N+1))
                self.boundary_conds = np.linspace(0, Smax, self.M+1)
        @abstractmethod
        def setup_boundary_conditions(self):
                raise NotImplementedError('Implementation required!')

        @abstractmethod
        def setup_coefficients(self):
                raise NotImplementedError('Implementation required!')

        @abstractmethod
        def traverse_grid(self):
                """ Iterate the grid backwards in time"""
                raise NotImplementedError('Implementation required!')

        @abstractmethod
        def interpolate(self):
                """ Use piecewise linear interpolation on the initial
                grid column to get the closest price at S0.
                """
                return np.interp(
                    self.S0, self.boundary_conds, self.grid[:,0])
```

Abstract base classes (ABCs) provide a way to define interfaces for a class. The `@abstractmethod()` decorator declares abstract methods that child classes should implement. Unlike Java's abstract methods, these methods may have an implementation and may be called via the `super()` mechanism from the class that overrides it.

In addition to these methods, we would need to define `dS` and `dt`, the change in `S` per unit time, and the change in `T` per iteration, respectively. We can write these as class properties:

```
@property
def dS(self):
    return self.Smax/float(self.M)

@property
```

```
def dt(self):
    return self.T/float(self.N)
```

Finally, add the `price()` method as the entry point that shows the steps in calling our discussed abstract methods:

```
def price(self):
    self.setup_boundary_conditions()
    self.setup_coefficients()
    self.traverse_grid()
    return self.interpolate()
```

A class for pricing European options using the explicit method of finite differences

The Python implementation of finite differences by using the explicit method is given in the following `FDExplicitEu` class, which inherits from the `FiniteDifferences` class and overrides the required implementation methods:

```
In [ ]:
    import numpy as np

    """
    Explicit method of Finite Differences
    """
    class FDExplicitEu(FiniteDifferences):

        def setup_boundary_conditions(self):
            if self.is_call:
                self.grid[:,-1] = np.maximum(
                    0, self.boundary_conds - self.K)
                self.grid[-1,:-1] = (self.Smax-self.K) * \
                    np.exp(-self.r*self.dt*(self.N-self.j_values))
            else:
                self.grid[:,-1] = np.maximum(
                    0, self.K-self.boundary_conds)
                self.grid[0,:-1] = (self.K-self.Smax) * \
                    np.exp(-self.r*self.dt*(self.N-self.j_values))

        def setup_coefficients(self):
            self.a = 0.5*self.dt*((self.sigma**2) *
                                  (self.i_values**2) -
                                  self.r*self.i_values)
            self.b = 1 - self.dt*((self.sigma**2) *
                                  (self.i_values**2) +
                                  self.r)
```

```
        self.c = 0.5*self.dt*((self.sigma**2) *
                              (self.i_values**2) +
                              self.r*self.i_values)

def traverse_grid(self):
    for j in reversed(self.j_values):
        for i in range(self.M)[2:]:
            self.grid[i,j] = \
                self.a[i]*self.grid[i-1,j+1] +\
                self.b[i]*self.grid[i,j+1]  + \
                self.c[i]*self.grid[i+1,j+1]
```

On completion of traversing the grid structure, the first column contains the present value of the initial asset prices at **t=0**. The `interp` function of NumPy is used to perform a linear interpolation to approximate the option value.

Besides using linear interpolation as the most common choice for the interpolation method, the other methods such as the spline or cubic may be used to approximate the option value.

Consider the example of a European put option. The underlying stock price is $50 with a volatility of 40 percent. The strike price of the put option is $50 with an expiration time of five months. The risk-free rate is 10 percent.

We can price this option using the explicit method with a `Smax` value of `100`, an `M` value of `100`, and an `N` value of `1000`:

```
In [ ]:
    option = FDExplicitEu(50, 50, r=0.1, T=5./12.,
        sigma=0.4, Smax=100, M=100, N=1000, is_put=True)
    print(option.price())
Out[ ]:
    4.072882278148043
```

What happens when other values of `M` and `N` are chosen improperly?

```
In [ ]:
    option = FDExplicitEu(50, 50, r=0.1, T=5./12.,
        sigma=0.4, Smax=100, M=80, N=100, is_put=True)
    print(option.price())
Out[ ]:
    -8.109445694129245e+35
```

It appears that the explicit method of the finite difference scheme suffers from instability problems.

The implicit method

The instability problem of the explicit method can be overcome using the forward difference with respect to time. The implicit method for approximating $f_{i,j}$ is given by the following equation:

$$rf_{i,j} = \frac{f_{i,j+1} - f_{i,j}}{dt} + ridS\frac{f_{i+1,j} - f_{i-1,j}}{2dS} + \frac{1}{2}\sigma^2 j^2 \frac{f_{i+1,j} - 2f_{i,j} + f_{i-1,j}}{dS^2}$$

Here, it can be seen that the only difference between the implicit and explicit approximating scheme lies in the first difference, where the forward difference with respect to t is used in the implicit scheme. When we rearrange the terms, we get the following expression:

$$f_{i,j+1} = a_j f_{i-1,j} + b_i f_{i,j} + c_i f_{i+1,j}$$

Where:

$$j = N - 1, N - 2, \ldots 2, 1, 0$$

$$i = 1, 2, 3, \ldots, M - 1$$

Here:

$$a_i = \frac{1}{2}(ridt - \sigma^2 i^2 dt)$$

$$b_i = 1 + \sigma^2 i^2 dt + rdt$$

$$c_i = -\frac{1}{2} + (ridt + \sigma^2 i^2 dt)$$

The iterative approach of the implicit scheme can be visually represented with the following diagram:

From the preceding diagram, we can note that values of *j+1* need to be computed before they can be used in the next iterative step, as the grid traverses backward. In the implicit scheme, the grid can be thought of as representing a system of linear equations at each iteration, as follows:

$$
\begin{bmatrix}
b_1 & c_1 & 0 & 0 & 0 & 0 \\
a_2 & b_2 & c_2 & 0 & 0 & 0 \\
0 & 0 & b_3 & \cdots & 0 & 0 \\
\vdots & \vdots & \vdots & \ddots & \vdots & \vdots \\
0 & 0 & 0 & a_{M-2} & b_{M-2} & c_{M-2} \\
0 & 0 & 0 & 0 & a_{M-1} & b_{M-1}
\end{bmatrix}
\begin{bmatrix}
f_{1,j} \\
f_{2,j} \\
f_{3,j} \\
\vdots \\
f_{M-2,j} \\
f_{M-1,j}
\end{bmatrix}
+
\begin{bmatrix}
a_1 f_{0,j} \\
0 \\
0 \\
\vdots \\
0 \\
C_{M-1} f_{M,j}
\end{bmatrix}
=
\begin{bmatrix}
f_{1,j+1} \\
f_{2,j+1} \\
f_{3,j+1} \\
\vdots \\
f_{M-2,j+1} \\
f_{M-1,j+1}
\end{bmatrix}
$$

By rearranging the terms, we get the following equation:

$$
\begin{bmatrix}
b_1 & c_1 & 0 & 0 & 0 & 0 \\
a_2 & b_2 & c_2 & 0 & 0 & 0 \\
0 & 0 & b_3 & \cdots & 0 & 0 \\
\vdots & \vdots & \vdots & \ddots & \vdots & \vdots \\
0 & 0 & 0 & a_{M-2} & b_{M-2} & c_{M-2} \\
0 & 0 & 0 & 0 & a_{M-1} & b_{M-1}
\end{bmatrix}
\begin{bmatrix}
f_{1,j} \\
f_{2,j} \\
f_{3,j} \\
\vdots \\
f_{M-2,j} \\
f_{M-1,j}
\end{bmatrix}
=
\begin{bmatrix}
f_{1,j+1} \\
f_{2,j+1} \\
f_{3,j+1} \\
\vdots \\
f_{M-2,j+1} \\
f_{M-1,j+1}
\end{bmatrix}
-
\begin{bmatrix}
a_1 f_{0,j} \\
0 \\
0 \\
\vdots \\
0 \\
C_{M-1} f_{M,j}
\end{bmatrix}
$$

The linear system of equations can be represented in the form of $Ax = B$, where we want to solve values of x in each iteration. Since the matrix A is tri-diagonal, we can use the LU factorization, where $A=LU$, for faster computation. Remember that we solved the linear system of equations using LU decomposition in `Chapter 2`, *The Importance of Linearity in Finance*.

A class for pricing European options using the implicit method of finite differences

The Python implementation of the implicit scheme is given in the following `FDImplicitEu` class. We can inherit the implementation of the explicit method from the `FDExplicitEu` class we discussed earlier and override the necessary methods of interest, namely, the `setup_coefficients` and `traverse_grid` methods:

```
In [ ]:
    import numpy as np
```

```
import scipy.linalg as linalg

"""
Explicit method of Finite Differences
"""
class FDImplicitEu(FDExplicitEu):

    def setup_coefficients(self):
        self.a = 0.5*(self.r*self.dt*self.i_values -
                      (self.sigma**2)*self.dt*\
                          (self.i_values**2))
        self.b = 1 + \
                 (self.sigma**2)*self.dt*\
                     (self.i_values**2) + \
                 self.r*self.dt
        self.c = -0.5*(self.r*self.dt*self.i_values +
                       (self.sigma**2)*self.dt*\
                           (self.i_values**2))
        self.coeffs = np.diag(self.a[2:self.M],-1) + \
                      np.diag(self.b[1:self.M]) + \
                      np.diag(self.c[1:self.M-1],1)

    def traverse_grid(self):
        """ Solve using linear systems of equations """
        P, L, U = linalg.lu(self.coeffs)
        aux = np.zeros(self.M-1)

        for j in reversed(range(self.N)):
            aux[0]  = np.dot(-self.a[1], self.grid[0, j])
            x1 = linalg.solve(L, self.grid[1:self.M, j+1]+aux)
            x2 = linalg.solve(U, x1)
            self.grid[1:self.M, j] = x2
```

Using the same example as the explicit scheme, we can price the European put options using the implicit scheme:

```
In [ ]:
    option = FDImplicitEu(50, 50, r=0.1, T=5./12.,
        sigma=0.4, Smax=100, M=100, N=1000, is_put=True)
    print(option.price())
Out[ ]:
    4.071594188049893
In [ ]:
    option = FDImplicitEu(50, 50, r=0.1, T=5./12.,
        sigma=0.4, Smax=100, M=80, N=100, is_put=True)
    print(option.price())
Out[ ]:
    4.063684691731647
```

Given the current parameters and input data, we can see that there are no stability issues with the implicit scheme.

The Crank-Nicolson method

Another way of avoiding the instability issue, as seen in the explicit method, is to use the Crank-Nicolson method. The Crank-Nicolson method converges much more quickly using a combination of the explicit and implicit methods, taking the average of both. This leads us to the following equation:

$$\frac{1}{2}rf_{i,j-1} + \frac{1}{2}rf_{i,j}$$

$$= \frac{f_{i,j} - f_{i,j-1}}{dt} \frac{1}{2}ridS\left(\frac{f_{i+1,j-1} - f_{i-1,j-1}}{2dS}\right) + \frac{1}{2}ridS\left(\frac{f_{i+1,j} - f_{i-1,j}}{2dS}\right)$$

$$+ \frac{1}{4}\sigma^2 i^2 dS^2\left(\frac{f_{i+1,j-1} - 2f_{i,j-1} + f_{i-1,j-1}}{dS^2}\right)$$

$$+ \frac{1}{4}\sigma^2 i^2 dS^2\left(\frac{f_{i+1,j} - 2f_{i,j} + f_{i-1,j}}{dS^2}\right)$$

This equation can also be rewritten as follows:

$$-\alpha_i f_{i-1,j-1} + (1 - \beta_i)f_{i,j-1} - \gamma_i f_{i+1,j-1} = \alpha_i f_{i-1,j} + (1 - \beta_i)f_{i,j-1} - \gamma_i f_{i+1,j}$$

Where:

$$\alpha_i = \frac{dt}{4}(\sigma^2 i^2 - ri)$$

$$\beta_i = \frac{dt}{2}(\sigma^2 i^2 + ri)$$

$$\gamma_i = \frac{dt}{4}(\sigma^2 i^2 + ri)$$

The iterative approach of the implicit scheme can be visually represented with the following diagram:

We can treat the equations as a system of linear equations in a matrix form:

$$M_1 f_{j-1} = M_2 f_j$$

Where:

$$M_1 = \begin{bmatrix} 1-\beta_1 & -\gamma_1 & 0 & 0 & 0 & 0 \\ -\alpha_2 & 1-\beta_2 & -\gamma_2 & 0 & 0 & 0 \\ 0 & -\alpha_3 & 1-\beta_3 & -\gamma_3 & 0 & 0 \\ 0 & 0 & \ddots & \ddots & \ddots & 0 \\ 0 & 0 & 0 & -\alpha_{M-2} & 1-\beta_{M-2} & \gamma_{M-2} \\ 0 & 0 & 0 & 0 & -\alpha_{M-1} & 1-\beta_{M-1} \end{bmatrix}$$

$$M_2 = \begin{bmatrix} 1+\beta_1 & \gamma_1 & 0 & 0 & 0 & 0 \\ \alpha_2 & 1+\beta_2 & \gamma_2 & 0 & 0 & 0 \\ 0 & \alpha_3 & 1+\beta_3 & -\gamma_3 & 0 & 0 \\ 0 & 0 & \ddots & \ddots & \ddots & 0 \\ 0 & 0 & 0 & \alpha_{M-2} & 1+\beta_{M-2} & \gamma_{M-2} \\ 0 & 0 & 0 & 0 & \alpha_{M-1} & 1+\beta_{M-1} \end{bmatrix}$$

$$f_i = [f_{1,j}, f_{2,j}, \ldots, f_{M-1,j}]^T$$

We can solve for the matrix M on every iterative procedure.

A class for pricing European options using the Crank-Nicolson method of finite differences

The Python implementation of the Crank-Nicolson method is given in the following FDCnEu class, which inherits from the FDExplicitEu class and overrides only the setup_coefficients and traverse_grid methods:

```
In [ ]:
    import numpy as np
    import scipy.linalg as linalg

    """
    Crank-Nicolson method of Finite Differences
    """
```

```
class FDCnEu(FDExplicitEu):

    def setup_coefficients(self):
        self.alpha = 0.25*self.dt*(
            (self.sigma**2)*(self.i_values**2) - \
            self.r*self.i_values)
        self.beta = -self.dt*0.5*(
            (self.sigma**2)*(self.i_values**2) + self.r)
        self.gamma = 0.25*self.dt*(
            (self.sigma**2)*(self.i_values**2) +
            self.r*self.i_values)
        self.M1 = -np.diag(self.alpha[2:self.M], -1) + \
                  np.diag(1-self.beta[1:self.M]) - \
                  np.diag(self.gamma[1:self.M-1], 1)
        self.M2 = np.diag(self.alpha[2:self.M], -1) + \
                  np.diag(1+self.beta[1:self.M]) + \
                  np.diag(self.gamma[1:self.M-1], 1)

    def traverse_grid(self):
        """ Solve using linear systems of equations """
        P, L, U = linalg.lu(self.M1)

        for j in reversed(range(self.N)):
            x1 = linalg.solve(
                L, np.dot(self.M2, self.grid[1:self.M, j+1]))
            x2 = linalg.solve(U, x1)
            self.grid[1:self.M, j] = x2
```

Using the same examples that we used with the explicit and implicit methods, we can price a European put option using the Crank-Nicolson method for different time point intervals:

```
In [ ]:
    option = FDCnEu(50, 50, r=0.1, T=5./12.,
        sigma=0.4, Smax=100, M=100, N=1000, is_put=True)
    print(option.price())
Out[ ]:
    4.072238354486825
In [ ]:
    option = FDCnEu(50, 50, r=0.1, T=5./12.,
        sigma=0.4, Smax=100, M=80, N=100, is_put=True)
    print(option.price())
Out[ ]:
    4.070145703042843
```

From the observed values, the Crank-Nicolson method not only avoids the instability issue we saw in the explicit scheme, but also converges faster than both the explicit and implicit methods. The implicit method requires more iterations, or bigger values of N, to produce values close to those of the Crank-Nicolson method.

Pricing exotic barrier options

Finite differences are especially useful in pricing exotic options. The nature of the option will dictate the specifications of the boundary conditions.

In this section, we will take a look at an example of pricing a down-and-out barrier option with the Crank-Nicolson method of finite differences. Due to its relative complexity, other analytical methods, such as Monte Carlo methods, are usually employed in favor of finite difference schemes.

A down-and-out option

Let's take a look at an example of a down-and-out option. At any time during the life of the option, should the underlying asset price fall below an $S_{barrier}$ barrier price, the option is considered worthless. Since, in the grid, the finite difference scheme represents all the possible price points, we only need to consider nodes with the following price range:

$$S_{barrier} \leq S_t \leq S_{max}$$

We can then set up the boundary conditions as follows:

$$f_{(S_{max}, t)} = 0$$
$$f_{(S_{barrier}, t)} = 0$$

A class for pricing down-and-out-options using the Crank-Nicolson method of finite differences

Let's create a class named FDCnDo that inherits from the FDCnEu class we discussed earlier. We will take into account the barrier price in the constructor method, while leaving the rest of the Crank-Nicolson implementation in the FDCnEu class unchanged:

```
In [ ]:
    import numpy as np

    """
    Price a down-and-out option by the Crank-Nicolson
    method of finite differences.
    """
    class FDCnDo(FDCnEu):

        def __init__(
            self, S0, K, r=0.05, T=1, sigma=0,
```

```
                    Sbarrier=0, Smax=1, M=1, N=1, is_put=False
        ):
            super(FDCnDo, self).__init__(
                S0, K, r=r, T=T, sigma=sigma,
                Smax=Smax, M=M, N=N, is_put=is_put
            )
            self.barrier = Sbarrier
            self.boundary_conds = np.linspace(Sbarrier, Smax, M+1)
            self.i_values = self.boundary_conds/self.dS

        @property
        def dS(self):
            return (self.Smax-self.barrier)/float(self.M)
```

Let's consider an example of a down-and-out option. The underlying stock price is $50 with a volatility of 40 percent. The strike price of the option is $50 with an expiration time of five months. The risk-free rate is 10 percent. The barrier price is $40.

We can price a call option and a put down-and-out option with Smax as 100, M as 120, and N as 500:

```
In [ ]:
    option = FDCnDo(50, 50, r=0.1, T=5./12.,
        sigma=0.4, Sbarrier=40, Smax=100, M=120, N=500)
    print(option.price())
Out [ ]:
    5.491560552934787
In [ ]:
    option = FDCnDo(50, 50, r=0.1, T=5./12., sigma=0.4,
        Sbarrier=40, Smax=100, M=120, N=500, is_put=True)
    print(option.price())
Out [ ]:
    0.5413635028954452
```

The prices of the down-and-out call and put options are $5.4916 and $0.5414, respectively.

Pricing American options with finite differences

So far, we have priced European options and exotic options. Due to the probability of an early exercise nature in American options, pricing such options is less straightforward. An iterative procedure is required in the implicit Crank-Nicolson method, where the payoffs from earlier exercises in the current period take into account the payoffs of an earlier exercise in the prior period. The Gauss-Siedel iterative method is proposed in the pricing of American options in the Crank- Nicolson method.

Recall that in Chapter 2, *The Importance of Linearity in Finance*, we covered the Gauss-Siedel method of solving systems of linear equations in the form of $Ax=B$. Here, the matrix A is decomposed into $A=L+U$, where L is a lower triangular matrix and U is an upper triangular matrix. Let's take a look at an example of a 4 x 4 matrix, A:

$$A = \begin{bmatrix} a & b & c & d \\ e & f & g & h \\ i & j & k & l \\ m & n & o & p \end{bmatrix} = \begin{bmatrix} a & 0 & 0 & 0 \\ e & f & 0 & 0 \\ i & j & k & 0 \\ m & n & o & p \end{bmatrix} + \begin{bmatrix} 0 & b & c & d \\ 0 & 0 & g & h \\ 0 & 0 & 0 & l \\ 0 & 0 & 0 & 0 \end{bmatrix}$$

The solution is then obtained iteratively, as follows:

$$Ax = B$$
$$(L+U)x = B$$
$$Lx = B - Ux$$
$$x_{n+1} = L^{-1}(B - U_x)$$

We can adapt the Gauss-Siedel method to our Crank-Nicolson implementation as follows:

$$r_j = M_1 f_{j-1} = M_2 f_j + \alpha_1 \begin{bmatrix} f_{0,j-1} + f_{0,j} \\ 0 \\ \vdots \\ 0 \end{bmatrix}$$

This equation satisfies the early exercise privilege equation:

$$f_{i,j-1} = max(f_{i,j-1}, K - idS)$$

A class for pricing American options using the Crank-Nicolson method of finite differences

Let's create a class named FDCnAm that inherits from the FDCnEu class, which is the Crank-Nicolson method's counterpart for pricing European options. The setup_coefficients method may be reused, while overriding all other methods for the inclusion of payoffs from an earlier exercise, if any.

The constructor __init__() and the setup_boundary_conditions() methods are given in the FDCnAm class:

```
In [ ]:
    import numpy as np
    import sys
```

```
"""
Price an American option by the Crank-Nicolson method
"""
class FDCnAm(FDCnEu):

    def __init__(self, S0, K, r=0.05, T=1,
            Smax=1, M=1, N=1, omega=1, tol=0, is_put=False):
        super(FDCnAm, self).__init__(S0, K, r=r, T=T,
            sigma=sigma, Smax=Smax, M=M, N=N, is_put=is_put)
        self.omega = omega
        self.tol = tol
        self.i_values = np.arange(self.M+1)
        self.j_values = np.arange(self.N+1)

    def setup_boundary_conditions(self):
        if self.is_call:
            self.payoffs = np.maximum(0,
                self.boundary_conds[1:self.M]-self.K)
        else:
            self.payoffs = np.maximum(0,
                self.K-self.boundary_conds[1:self.M])

        self.past_values = self.payoffs
        self.boundary_values = self.K * np.exp(
                -self.r*self.dt*(self.N-self.j_values))
```

Next, implement the `traverse_grid()` method in the same class:

```
def traverse_grid(self):
    """ Solve using linear systems of equations """
    aux = np.zeros(self.M-1)
    new_values = np.zeros(self.M-1)

    for j in reversed(range(self.N)):
        aux[0] = self.alpha[1]*(self.boundary_values[j] +
                            self.boundary_values[j+1])
        rhs = np.dot(self.M2, self.past_values) + aux
        old_values = np.copy(self.past_values)
        error = sys.float_info.max

        while self.tol < error:
            new_values[0] = \
                self.calculate_payoff_start_boundary(
                    rhs, old_values)

            for k in range(self.M-2)[1:]:
                new_values[k] = \
                    self.calculate_payoff(
```

```
                   k, rhs, old_values, new_values)

         new_values[-1] = \
             self.calculate_payoff_end_boundary(
                 rhs, old_values, new_values)

         error = np.linalg.norm(new_values-old_values)
         old_values = np.copy(new_values)

     self.past_values = np.copy(new_values)

 self.values = np.concatenate(
     ([self.boundary_values[0]], new_values, [0]))
```

In each iterative procedure of the `while` loop, the payoffs are calculated while taking into account the start and end boundaries. Furthermore, `new_values` are constantly replaced with new payoff calculations based on existing and previous values.

At the start boundaries where the index is 0, the payoffs are calculated with the alpha values omitted. Implement the `calculate_payoff_start_boundary()` method inside the class:

```
def calculate_payoff_start_boundary(self, rhs, old_values):
    payoff = old_values[0] + \
        self.omega/(1-self.beta[1]) * \
            (rhs[0] - \
            (1-self.beta[1])*old_values[0] + \
            self.gamma[1]*old_values[1])

    return max(self.payoffs[0], payoff)
```

At the end boundary where the last index is, the payoffs are calculated with the gamma values omitted. Implement the `calculate_payoff_end_boundary()` method inside the class:

```
def calculate_payoff_end_boundary(self, rhs, old_values, new_values):
    payoff = old_values[-1] + \
        self.omega/(1-self.beta[-2]) * \
            (rhs[-1] + \
            self.alpha[-2]*new_values[-2] - \
            (1-self.beta[-2])*old_values[-1])

    return max(self.payoffs[-1], payoff)
```

For payoffs that are not at the boundaries, the payoffs are calculated by taking into account the alpha and gamma values. Implement the `calculate_payoff()` method inside the class:

```
def calculate_payoff(self, k, rhs, old_values, new_values):
    payoff = old_values[k] + \
        self.omega/(1-self.beta[k+1]) * \
            (rhs[k] + \
            self.alpha[k+1]*new_values[k-1] - \
            (1-self.beta[k+1])*old_values[k] + \
            self.gamma[k+1]*old_values[k+1])

    return max(self.payoffs[k], payoff)
```

Since the new variable, `values`, contains our terminal payoff values as a one-dimensional array, override the parent `interpolate` method to account for this change with the following code:

```
def interpolate(self):
    # Use linear interpolation on final values as 1D array
    return np.interp(self.S0, self.boundary_conds, self.values)
```

The tolerance parameter is used in the Gauss-Siedel method as the convergence criterion. The `omega` variable is the over-relaxation parameter. Higher `omega` values provide faster convergence, but this also comes with higher possibilities of the algorithm not converging.

Let's price an American call-and-put option with an underlying asset price of 50 and volatility of 40 percent, a strike price of 50, a risk-free rate of 10 percent, and an expiration date of five months. We choose a `Smax` value of `100`, M as `100`, N as `42`, an `omega` parameter value of `1.2`, and a tolerance value of `0.001`:

```
In [ ]:
    option = FDCnAm(50, 50, r=0.1, T=5./12.,
        sigma=0.4, Smax=100, M=100, N=42, omega=1.2, tol=0.001)
    print(option.price())
Out[ ]:
    6.108682815392217
In [ ]:
    option = FDCnAm(50, 50, r=0.1, T=5./12., sigma=0.4, Smax=100,
        M=100, N=42, omega=1.2, tol=0.001, is_put=True)
    print(option.price())
Out[ ]:
    4.277764229383736
```

The prices of the call and put American stock options by using the Crank-Nicolson method are \$6.109 and \$4.2778, respectively.

Putting it all together – implied volatility modeling

In the option pricing methods we have learned so far, a number of parameters are assumed to be constant: interest rates, strike prices, dividends, and volatility. Here, the parameter of interest is volatility. In quantitative research, the volatility ratio is used to forecast price trends.

To derive implied volatilities, we need to refer to `Chapter 3`, *Nonlinearity in Finance*, where we discussed the root-finding methods of nonlinear functions. We will use the bisection method of numerical procedures in our next example to create an implied volatility curve.

Implied volatilities of the AAPL American put option

Let's consider the option data of the stock **Apple** (**AAPL**), which was gathered at the end of the day on October 3, 2014. These details are provided in the following table. The option expires on December 20, 2014. The prices listed are the mid-points of the bid and ask prices:

Strike price	Call price	Put price
75	30	0.16
80	24.55	0.32
85	20.1	0.6
90	15.37	1.22
92.5	10.7	1.77
95	8.9	2.54
97.5	6.95	3.55
100	5.4	4.8
105	4.1	7.75
110	2.18	11.8
115	1.05	15.96
120	0.5	20.75
125	0.26	25.8

The last traded price of AAPL was 99.62, with an interest rate of 2.48 percent and a dividend yield of 1.82 percent. The American options expire in 78 days.

Using this information, let's create a new class named `ImpliedVolatilityModel` that accepts the stock option's parameters in the constructor. If required, import the `BinomialLROption` class that we created for the LR binomial tree we covered in the earlier section of this chapter, *A class for the LR binomial tree option pricing model*. The `bisection` function we covered in `Chapter 3`, *Nonlinearity in Finance*, is also required.

The `option_valuation()` method accepts the K strike price and the `sigma` volatility value to compute the value of the option. In this example, we are using the `BinomialLROption` pricing method.

The `get_implied_volatilities()` method accepts a list of strike and option prices to compute the implied volatilities by the `bisection` method for every price available. Therefore, the length of the two lists must be the same.

The Python code for the `ImpliedVolatilityModel` class is given as follows:

```
In [ ]:
    """
    Get implied volatilities from a Leisen-Reimer binomial
    tree using the bisection method as the numerical procedure.
    """
    class ImpliedVolatilityModel(object):

        def __init__(self, S0, r=0.05, T=1, div=0,
                     N=1, is_put=False):
            self.S0 = S0
            self.r = r
            self.T = T
            self.div = div
            self.N = N
            self.is_put = is_put

        def option_valuation(self, K, sigma):
            """ Use the binomial Leisen-Reimer tree """
            lr_option = BinomialLROption(
                self.S0, K, r=self.r, T=self.T, N=self.N,
                sigma=sigma, div=self.div, is_put=self.is_put
            )
            return lr_option.price()

        def get_implied_volatilities(self, Ks, opt_prices):
            impvols = []
            for i in range(len(strikes)):
                # Bind f(sigma) for use by the bisection method
                f = lambda sigma: \
                    self.option_valuation(Ks[i], sigma)-\
```

```
            opt_prices[i]
        impv = bisection(f, 0.01, 0.99, 0.0001, 100)[0]
        impvols.append(impv)
    return impvols
```

Import the `bisection` function we discussed in the previous chapter:

```
In [ ]:
    def bisection(f, a, b, tol=0.1, maxiter=10):
        """
        :param f: The function to solve
        :param a: The x-axis value where f(a)<0
        :param b: The x-axis value where f(b)>0
        :param tol: The precision of the solution
        :param maxiter: Maximum number of iterations
        :return: The x-axis value of the root,
                 number of iterations used
        """
        c = (a+b)*0.5  # Declare c as the midpoint ab
        n = 1  # Start with 1 iteration
        while n <= maxiter:
            c = (a+b)*0.5
            if f(c) == 0 or abs(a-b)*0.5 < tol:
                # Root is found or is very close
                return c, n

            n += 1
            if f(c) < 0:
                a = c
            else:
                b = c

        return c, n
```

Using this model, let's find out the implied volatilities of the American put options using this particular set of data:

```
In [ ]:
    strikes = [75, 80, 85, 90, 92.5, 95, 97.5,
               100, 105, 110, 115, 120, 125]
    put_prices = [0.16, 0.32, 0.6, 1.22, 1.77, 2.54, 3.55,
                  4.8, 7.75, 11.8, 15.96, 20.75, 25.81]
In [ ]:
    model = ImpliedVolatilityModel(
        99.62, r=0.0248, T=78/365., div=0.0182, N=77, is_put=True)
    impvols_put = model.get_implied_volatilities(strikes, put_prices)
```

The implied volatility values are now stored in the `impvols_put` variable as a `list` object. Let's plot these values against the strike prices to obtain an implied volatility curve:

```
In [ ]:
    %matplotlib inline
    import matplotlib.pyplot as plt

    plt.plot(strikes, impvols_put)
    plt.xlabel('Strike Prices')
    plt.ylabel('Implied Volatilities')
    plt.title('AAPL Put Implied Volatilities expiring in 78 days')
    plt.show()
```

This would give us the volatility smile, as shown in the following diagram. Here, we have modeled an LR tree with 77 steps, with each step representing one day:

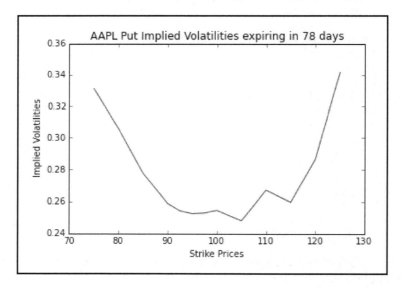

Of course, pricing an option daily may not be ideal since markets change by fractions of a millisecond. We used the bisection method to solve the implied volatility as implied by the binomial tree, as opposed to the realized volatility values directly observed from market prices.

Should we fit this curve against a polynomial curve to identify potential arbitrage opportunities? Or extrapolate the curve to derive further insights on potential opportunities from implied volatilities of far out-of-the-money and in-the-money options? Well, these questions are for option traders like yourself to find out!

Summary

In this chapter, we looked at a number of numerical procedures in derivative pricing, the most common being options. One such procedure is the use of trees, with binomial trees being the simplest structure to model asset information, where one node extends to two other nodes in each time step, representing an up state and a down state, respectively. In trinomial trees, each node extends to three other nodes in each time step, representing an up state, a down state, and a state with no movement, respectively. As the tree traverses upwards, the underlying asset is computed and represented at each node. The option then takes on the structure of this tree and, starting from the terminal payoffs, the tree traverses backward and toward the root, which converges to the current discounted option price. Besides binomial and trinomial trees, trees can take on the form of the CRR, Jarrow-Rudd, Tian, or LR parameters.

By adding another layer of nodes around our tree, we introduced additional information from which we can derive the Greeks, such as the delta and gamma, without incurring additional computational costs.

Lattices were introduced as a way of saving storage costs over binomial and trinomial trees. In lattice pricing, nodes with new information are saved only once and reused later on nodes that require no change in the information.

We also discussed the finite difference schemes in option pricing, consisting of terminal and boundary conditions. From the terminal conditions, the grid traverses backward in time using the explicit method, implicit method, and the Crank- Nicolson method. Besides pricing European and American options, finite difference pricing schemes can be used to price exotic options, where we looked at an example of pricing a down-and-out barrier option.

By importing the bisection root-finding method learned about in `Chapter 3`, *Nonlinearity in Finance*, and the binomial LR tree model in this chapter, we used market prices of an American option to create an implied volatility curve for further studies.

In the next chapter, we will take a look at modeling interest rates and derivatives.

5

Modeling Interest Rates and Derivatives

Interest rates affect economic activities at all levels. Central banks, including the **Federal Reserve** (informally known as the **Fed**), target interest rates as a policy tool to influence economic activity. Interest rate derivatives are popular with investors who require customized cash flow needs or specific views on interest-rate movements.

One of the key challenges that interest-rate derivative traders face is to have a good and robust pricing procedure for these products. This involves understanding the complicated behavior of an individual interest-rate movement. Several interest-rate models have been proposed for financial studies. Some common models studied in finance are the Vasicek, CIR, and Hull-White models. These interest-rate models involve modeling the short-rate and rely on factors (or sources of uncertainty) with most of them using only one factor. Two-factor and multi-factor interest rate models have been proposed.

In this chapter, we will cover the following topics:

- Understanding yield curves
- Valuing a zero-coupon bond
- Bootstrapping a yield curve
- Calculating forward rates
- Calculating the yield to maturity and price of a bond
- Calculating the bond duration and convexity using Python
- Short-rate modeling
- The Vasicek short-rate model
- Types of bond options
- Pricing a callable bond option

Fixed-income securities

Corporations and governments issue fixed-income securities as a means of raising money. The owners of such debts lend money and expect to receive the principal when the debt matures. The issuer who wishes to borrow money may issue a fixed amount interest payment during the lifetime of the debt at pre-specified times.

The holders of debt securities, such as US Treasury bills, notes, and bonds, face the risk of default by the issuer. The federal government and municipal government are thought to face the least default risk, since they can easily raise taxes and create more money to repay the outstanding debts.

Most bonds pay a fixed amount of interest semi-annually, while some pay quarterly, or annually. These interest payments are also referred to as coupons. They are quoted as a percentage of the face value or par amount of the bond on an annual basis.

For example, a 5-year $10,000 Treasury bond with a coupon rate of 5% pays coupons of $500 each year, or coupons of $250 every 6 months, up to and including the maturity date. Should the interest rates drop and new T-bonds pay a 3% coupon rate, the buyer of the new bond will only receive coupons of $300 annually, while existing holders of the 5% bond will continue to receive $500 annually. As the characteristics of the bonds influence their prices, they're closely related to current levels of interest rates in an inverse way: the value of the bond decreases as the interest rates increase. As interest rates decrease, bond prices increase.

Yield curves

In a normal yield curve environment, long-term interest rates are higher than short-term interest rates. Investors expect to be compensated with higher returns when they lend money for a longer period since they are exposed to a higher default risk. The normal or positive yield curve is said to be upward sloping, as shown in the following graph:

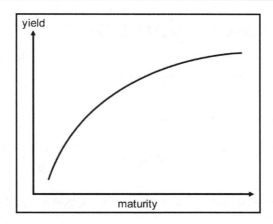

In certain economic conditions, the yield curve can be inverted. Long-term interest rates are lower than short-term interest rates. Such a condition occurs when the supply of money is tight. Investors are willing to forgo long-term gains to preserve their wealth in the short-term. During periods of high inflation, where the inflation rate exceeds the rate of coupon interests, negative interest rates may be observed. Investors are willing to pay in the short-term just to secure their long-term wealth. The inverted yield curve is said to be downward sloping, as shown in the following graph:

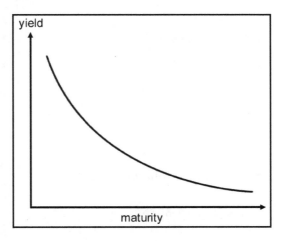

Valuing a zero-coupon bond

A **zero-coupon bond** is a bond that does not pay any periodic interest except on maturity, where the principal or face value is repaid. Zero-coupon bonds are also called **pure discount bonds**.

A zero-coupon bond can be valued as follows:

$$\text{price of zero-coupon bond} = \frac{face\ value}{(1 + y)^t}$$

Here, y is the annually-compounded yield or rate of the bond, and t is the time remaining to the maturity of the bond.

Let's take a look at an example of a five-year zero-coupon bond with a face value of \$100. The yield is 5%, compounded annually. The price can be calculated as follows:

$$\frac{100}{(1 + 0.05)^5} = \$78.35$$

A simple Python zero-coupon bond calculator can be used to illustrate this example:

```
In [ ]:
    def zero_coupon_bond(par, y, t):
        """
        Price a zero coupon bond.

        :param par: face value of the bond.
        :param y: annual yield or rate of the bond.
        :param t: time to maturity, in years.
        """
        return par/(1+y)**t
```

Using the preceding example, we get the following result:

```
In [ ]:
    print(zero_coupon_bond(100, 0.05, 5))
Out[ ]:
    78.35261664684589
```

In the preceding example, we assumed that the investor is able to invest \$78.35 at the prevailing annual interest rate of 5% for 5 years, compounded annually.

Now that we have a zero-coupon bond calculator, we can use it to determine zero rates by bootstrapping the yield curve, as explained in the next section.

Spot and zero rates

As the compounding frequency increases (say, from compounded yearly to compounded daily), the future value of money reaches an exponential limit. That is to say, the value of $100 today will reach a future value of $100e^{RT}$ when it is invested at a continuously compounded rate, R, for a period of time, T. If we discount these values for a security that pays $100 at a future time, T, with a continuously-compounded discount rate, R, its value at time zero is $\frac{100}{e^{RT}}$. This rate is known as the **spot rate**.

Spot rates represent the current interest rates for several maturities, should we want to borrow or lend money now. Zero rates represent the internal rate of return of zero-coupon bonds.

By deriving the spot rates of bonds with different maturities, we can construct the present yield curve through a bootstrapping process using zero-coupon bonds.

Bootstrapping a yield curve

Short-term spot rates can be derived directly from various short-term securities, such as zero-coupon bonds, T-bills, notes, and eurodollar deposits. However, longer-term spot rates are typically derived from the prices of long-term bonds through a bootstrapping process, taking into account the spot rates of maturities that correspond to the coupon payment date. After obtaining short-term and long-term spot rates, the yield curve can then be constructed.

An example of bootstrapping the yield curve

Let's illustrate the bootstrapping of the yield curve with an example. The following table shows a list of bonds with different maturities and prices:

Bond face value in dollars	Time to maturity in years	Annual coupon in dollars	Bond cash price in dollars
100	0.25	0	97.50
100	0.50	0	94.90
100	1.00	0	90.00
100	1.50	8	96.00
100	2.00	12	101.60

An investor in a three-month zero-coupon bond today at \$97.50 would earn interest of \$2.50. The three-month spot rate can be calculated as follows:

$$97.50 = \frac{100}{e^{0.25y}}$$

$$e^{0.25y} = 1.0256$$

$$y = 4\ln 1.0256 = 0.10127$$

Thus, the 3-month zero rate is 10.127% with continuous compounding. The spot rates of the zero-coupon bonds are shown in the following table:

Time to maturity in years	Spot rate (in percent)
0.25	10.127
0.50	10.469
1.00	10.536

Using these spot rates, we can now price the 1.5-year bond as follows:

$$4e^{(-0.10469)(0.5)} + 4e^{(-0.10536)(1.0)} + 104e^{(-y)(1.5)} = 96$$

The value of y can easily be solved by rearranging the equation, as follows:

$$y = -ln\left(\frac{96 - 4e^{(-0.10469)(0.5)} - 4e^{(-0.10536)(1.0)}}{104}\right) \div 1.5 = 0.106809$$

With the spot rate of the 1.5-year bond as 10.681%, we can use it to price the 2-year bond with coupons of \$6 semi-annually, as follows:

$$6e^{(-0.10469)(0.5)} + 6e^{(-0.10536)(1)} + 6e^{(-0.106809)(1.5)} + 106e^{(-y)(2)} = 101.60$$

Rearranging the equation and solving for y gives us the spot rate of the 2-year bond as 10.808.

Through this iterative procedure of calculating the spot rate of every bond in increasing order of maturity and using it in the next iteration, we obtain a list of spot rates of different maturities that we can use to construct the yield curve.

Writing the yield curve bootstrapping class

The steps for writing the Python code to bootstrap the yield curve and running with a plot output are outlined as follows:

1. Create a class named `BootstrapYieldCurve`, which will implement the bootstrapping of the yield curve in Python code:

   ```python
   import math

   class BootstrapYieldCurve(object):

       def __init__(self):
           self.zero_rates = dict()
           self.instruments = dict()
   ```

2. In the constructor, the two `zero_rates` and `instruments` dictionary variables are declared and will be used by several methods, as follows:

   - Add a method named `add_instrument()`, which appends a tuple of bond information to the `instruments` dictionary, indexed by the maturity time. This method is written as follows:

     ```python
     def add_instrument(self, par, T, coup, price,
     compounding_freq=2):
         self.instruments[T] = (par, coup, price, compounding_freq)
     ```

   - Add a method named `get_maturities()`, which simply returns a list of available maturities in ascending order. This method is written as follows:

     ```python
     def get_maturities(self):
         """
         :return: a list of maturities of added instruments
         """
         return sorted(self.instruments.keys())
     ```

- Add a method named `get_zero_rates()`, which bootstraps the yield curve, calculates the spot rates along that yield curve, and returns a list of zero rates in ascending order of maturity. This method is written as follows:

```python
def get_zero_rates(self):
    """
    Returns a list of spot rates on the yield curve.
    """
    self.bootstrap_zero_coupons()
    self.get_bond_spot_rates()
    return [self.zero_rates[T] for T in self.get_maturities()]
```

- Add a method named `bootstrap_zero_coupons()`, which calculates the spot rates of the given zero-coupon bonds and adds them to the `zero_rates` dictionary, indexed by maturity. This method is written as follows:

```python
def bootstrap_zero_coupons(self):
    """
    Bootstrap the yield curve with zero coupon instruments
    first.
    """
    for (T, instrument) in self.instruments.items():
        (par, coup, price, freq) = instrument
        if coup == 0:
            spot_rate = self.zero_coupon_spot_rate(par, price,
T)
            self.zero_rates[T] = spot_rate
```

- Add a method named `zero_coupon_spot_rate()`, which calculates the spot rate of a zero-coupon bond. This method is called by `bootstrap_zero_coupons()` and is written as follows:

```python
def zero_coupon_spot_rate(self, par, price, T):
    """
    :return: the zero coupon spot rate with continuous
compounding.
    """
    spot_rate = math.log(par/price)/T
    return spot_rate
```

- Add a method named `get_bond_spot_rates()`, which calculates the spot rates of non zero-coupon bonds and adds them to the `zero_rates` dictionary, indexed by maturity. This method is written as follows:

```
def get_bond_spot_rates(self):
    """
    Get spot rates implied by bonds, using short-term
    instruments.
    """
    for T in self.get_maturities():
        instrument = self.instruments[T]
        (par, coup, price, freq) = instrument
        if coup != 0:
            spot_rate = self.calculate_bond_spot_rate(T,
instrument)
            self.zero_rates[T] = spot_rate
```

- Add a method named `calculate_bond_spot_rate()`, which is called by `get_bond_spot_rates()` to calculate the spot rate at a particular maturity period. This method is written as follows:

```
def calculate_bond_spot_rate(self, T, instrument):
    try:
        (par, coup, price, freq) = instrument
        periods = T*freq
        value = price
        per_coupon = coup/freq
        for i in range(int(periods)-1):
            t = (i+1)/float(freq)
            spot_rate = self.zero_rates[t]
            discounted_coupon = per_coupon*math.exp(-
spot_rate*t)
            value -= discounted_coupon

        last_period = int(periods)/float(freq)
        spot_rate = -
math.log(value/(par+per_coupon))/last_period
        return spot_rate
    except:
        print("Error: spot rate not found for T=", t)
```

3. Instantiate the `BootstrapYieldCurve` class and add each bond's information from the preceding table:

```
In [ ]:
    yield_curve = BootstrapYieldCurve()
    yield_curve.add_instrument(100, 0.25, 0., 97.5)
    yield_curve.add_instrument(100, 0.5, 0., 94.9)
    yield_curve.add_instrument(100, 1.0, 0., 90.)
    yield_curve.add_instrument(100, 1.5, 8, 96., 2)
    yield_curve.add_instrument(100, 2., 12, 101.6, 2)
In [ ]:
    y = yield_curve.get_zero_rates()
    x = yield_curve.get_maturities()
```

4. Calling the `get_zero_rates()` method in the class instance returns a list of spot rates in the same order as the maturities, which are stored in the x and y variables, respectively. Issue the following Python code to plot x and y on a graph:

```
In [ ]:
    %pylab inline

    fig = plt.figure(figsize=(12, 8))
    plot(x, y)
    title("Zero Curve")
    ylabel("Zero Rate (%)")
    xlabel("Maturity in Years");
```

5. We get the following yield curve:

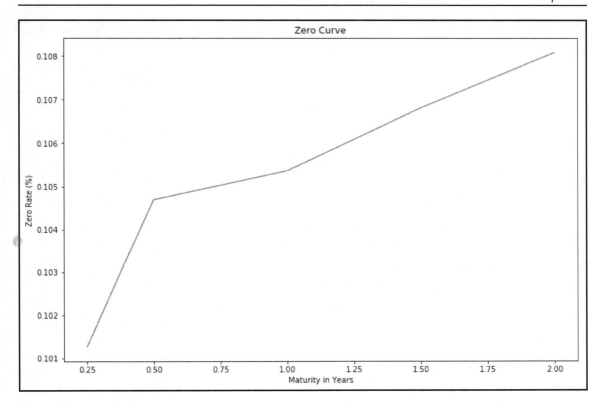

In a normal yield curve environment, where the interest rates increase as the maturities increase, we obtain an upward-sloping yield curve.

Forward rates

An investor who plans to invest at a later time may be curious to know what the future interest rate will look like, as implied by today's term structure of interest rates. For example, you might ask, *What is the one-year spot rate one year from now?* To answer this question, you can calculate forward rates for the period between T_1 and T_2 using this formula:

$$r_{forward} = \frac{r_2 T_2 - r_1 T_1}{T_2 - T_1}$$

Here, r_1 and r_2 are the continuously-compounded annual interest rates at time periods T_1 and T_2, respectively.

The following `ForwardRates` class helps us generate a list of forward rates from a list of spot rates:

```
class ForwardRates(object):

    def __init__(self):
        self.forward_rates = []
        self.spot_rates = dict()

    def add_spot_rate(self, T, spot_rate):
        self.spot_rates[T] = spot_rate

    def get_forward_rates(self):
        """
        Returns a list of forward rates
        starting from the second time period.
        """
        periods = sorted(self.spot_rates.keys())
        for T2, T1 in zip(periods, periods[1:]):
            forward_rate = self.calculate_forward_rate(T1, T2)
            self.forward_rates.append(forward_rate)

        return self.forward_rates

    def calculate_forward_rate(self, T1, T2):
        R1 = self.spot_rates[T1]
        R2 = self.spot_rates[T2]
        forward_rate = (R2*T2-R1*T1)/(T2-T1)
        return forward_rate
```

Using spot rates derived from our preceding yield curve, we get the following result:

```
In [ ]:
    fr = ForwardRates()
    fr.add_spot_rate(0.25, 10.127)
    fr.add_spot_rate(0.50, 10.469)
    fr.add_spot_rate(1.00, 10.536)
    fr.add_spot_rate(1.50, 10.681)
    fr.add_spot_rate(2.00, 10.808)
In [ ]:
    print(fr.get_forward_rates())
Out[ ]:
    [10.810999999999998, 10.603, 10.971, 11.189]
```

Calling the `get_forward_rates()` method of the `ForwardRates` class returns a list of forward rates, starting from the next time period.

Calculating the yield to maturity

The **yield to maturity** (**YTM**) measures the interest rate, as implied by the bond, which takes into account the present value of all the future coupon payments and the principal. It is assumed that bond holders can invest received coupons at the YTM rate until the maturity of the bond; according to risk-neutral expectations, the payments received should be the same as the price paid for the bond.

Let's take a look at an example of a 5.75% bond that will mature in 1.5 years with a par value of 100. The price of the bond is $95.0428 and coupons are paid semi-annually. The pricing equation can be stated as follows:

$$95.0428 = \frac{c}{\left(1 + \frac{y}{n}\right)^{nT_1}} + \frac{c}{\left(1 + \frac{y}{n}\right)^{nT_2}} + \frac{100 + c}{\left(1 + \frac{y}{n}\right)^{nT_3}}$$

Here:

- c is the coupon dollar amount paid at each time period
- T is the time period of payment in years
- n is the coupon payment frequency
- y is the YTM that we are interested in solving

To solve the YTM is typically a complex process, and most bond YTM calculators use Newton's method as an iterative process.

The bond YTM calculator is illustrated by the following `bond_ytm()` function:

```
import scipy.optimize as optimize

def bond_ytm(price, par, T, coup, freq=2, guess=0.05):
    freq = float(freq)
    periods = T*2
    coupon = coup/100.*par
    dt = [(i+1)/freq for i in range(int(periods))]
    ytm_func = lambda y: \
        sum([coupon/freq/(1+y/freq)**(freq*t) for t in dt]) +\
        par/(1+y/freq)**(freq*T) - price

    return optimize.newton(ytm_func, guess)
```

Remember that we covered the use of Newton's method and other nonlinear function root solvers in `Chapter 3`, *Nonlinearity in Finance*. For this YTM calculator function, we used the `scipy.optimize` package to solve the YTM.

Using the parameters from the bond example, we get the following result:

```
In [ ] :
    ytm = bond_ytm(95.0428, 100, 1.5, 5.75, 2)
In [ ]:
    print(ytm)
Out[ ]:
    0.09369155345239522
```

The YTM of the bond is 9.369%. Now we have a bond YTM calculator that can help us compare a bond's expected return with those of other securities.

Calculating the price of a bond

When the YTM is known, we can get back the bond price in the same way we used the pricing equation. This is implemented by the `bond_price()` function:

```
In [ ]:
    def bond_price(par, T, ytm, coup, freq=2):
        freq = float(freq)
        periods = T*2
        coupon = coup/100.*par
        dt = [(i+1)/freq for i in range(int(periods))]
        price = sum([coupon/freq/(1+ytm/freq)**(freq*t) for t in dt]) + \
            par/(1+ytm/freq)**(freq*T)
        return price
```

Plugging in the same values from the earlier example, we get the following result:

```
In [ ]:
    price = bond_price(100, 1.5, ytm, 5.75, 2)
    print(price)
Out[ ]:
    95.04279999999997
```

This gives us the same original bond price discussed in the earlier example, *Calculating the yield to maturity*. With the `bond_ytm()` and `bond_price()` functions, we can apply these for further uses in bond pricing, such as finding the bond's modified duration and convexity. These two characteristics of bonds are of importance to bond traders in helping them formulate various trading strategies and hedge their risk.

Bond duration

Duration is a sensitivity measure of bond prices to yield changes. Some duration measures are effective duration, Macaulay duration, and modified duration. The type of duration that we will discuss is modified duration, which measures the percentage change in bond price with respect to a percentage change in yield (typically 1% or 100 **basis points (bps)**).

The higher the duration of a bond, the more sensitive it is to yield changes. Conversely, the lower the duration of a bond, the less sensitive it is to yield changes.

The modified duration of a bond can be thought of as the first derivative of the relationship between price and yield:

$$modified\ duration \approx \frac{P^- - P^+}{2(P_0)(dY)}$$

Here:

- dY is the given change in yield
- P^- is the price of the bond from a decrease in yield by dY
- P^+ is the price of the bond from an increase in yield by dY
- P_0 is the initial price of the bond

It should be noted that the duration describes the linear price-yield relationship for a small change in Y. Because the yield curve is not linear, using a large value of dY does not approximate the duration measure well.

The implementation of the modified duration calculator is given in the following `bond_mod_duration()` function. It uses the `bond_ytm()` function as discussed earlier in this chapter, *Calculating the yield to maturity*, to determine the yield of the bond with the given initial value. Also, it uses the `bond_price()` function to determine the price of the bond with the given change in yield:

```
In [ ]:
    def bond_mod_duration(price, par, T, coup, freq, dy=0.01):
        ytm = bond_ytm(price, par, T, coup, freq)
        ytm_minus = ytm - dy
        price_minus = bond_price(par, T, ytm_minus, coup, freq)
        ytm_plus = ytm + dy
        price_plus = bond_price(par, T, ytm_plus, coup, freq)
        mduration = (price_minus-price_plus)/(2*price*dy)
        return mduration
```

We can find out the modified duration of the 5.75% bond discussed earlier, in *Calculating the yield to maturity*, which will mature in 1.5 years with a par value of 100 and a bond price of 95.0428:

```
In [ ]:
    mod_duration = bond_mod_duration(95.0428, 100, 1.5, 5.75, 2)
In [ ]:
    print(mod_duration)
Out[ ]:
    1.3921935426561034
```

The modified duration of the bond is 1.392 years.

Bond convexity

Convexity is the sensitivity measure of the duration of a bond to yield changes. Think of convexity as the second derivative of the relationship between the price and yield:

$$convexity \approx \frac{P^- + P^+ - 2P_0}{(P_0)(dY)^2}$$

Bond traders use convexity as a risk-management tool to measure the amount of market risk in their portfolio. Higher-convexity portfolios are less affected by interest-rate volatilities than lower-convexity portfolios, given the same bond duration and yield. As such, higher-convexity bonds are more expensive than lower-convexity ones, everything else being equal.

The implementation of a bond convexity is given as follows:

```
In [ ]:
    def bond_convexity(price, par, T, coup, freq, dy=0.01):
        ytm = bond_ytm(price, par, T, coup, freq)
        ytm_minus = ytm - dy
        price_minus = bond_price(par, T, ytm_minus, coup, freq)
        ytm_plus = ytm + dy
        price_plus = bond_price(par, T, ytm_plus, coup, freq)
        convexity = (price_minus + price_plus - 2*price)/(price*dy**2)
        return convexity
```

We can now find the convexity of the 5.75% bond discussed earlier, in *Calculating the yield to maturity* section, which will mature in 1.5 years with a par value of 100 and a bond price of 95.0428:

```
In [ ]:
    convexity = bond_convexity(95.0428, 100, 1.5, 5.75, 2)
In [ ]:
    print(convexity)
Out[ ] :
    2.633959390331875
```

The convexity of the bond is 2.63. For two bonds with the same par value, coupon, and maturity, their convexity may be different, depending on their location on the yield curve. Higher-convexity bonds will exhibit higher price changes for the same change in yield.

Short–rate modeling

In short-rate modeling, the short-rate, *r(t)*, is the spot rate at a particular time. It is described as a continuously-compounded, annualized interest rate term for an infinitesimally short period of time on the yield curve. The short-rate takes on the form of a stochastic variable in interest-rate models, where the interest rates may change by small amounts at every point in time. Short-rate models attempt to model the evolution of interest rates over time, and hopefully describe the economic conditions at certain periods.

Short-rate models are frequently used in the evaluation of interest-rate derivatives. Bonds, credit instruments, mortgages, and loan products are sensitive to interest-rate changes. Short-rate models are used as interest rate components in conjunction with pricing implementations, such as numerical methods, to help price such derivatives.

Interest-rate modeling is considered a fairly complex topic since interest rates are affected by a multitude of factors, such as economic states, political decisions, government intervention, and laws of supply and demand . A number of interest-rate models have been proposed to account for various characteristics of interest rates.

In this section, we will take a look at some of the most popular one-factor short-rate models used in financial studies, namely, the Vasicek, Cox-Ingersoll-Ross, Rendleman and Bartter, and Brennan and Schwartz models. Using Python, we will perform a one-path simulation to obtain a general overview of the interest-rate path process. Other models commonly discussed in finance include the Ho-Lee, Hull-White, and Black-Karasinki.

The Vasicek model

In the one-factor Vasicek model, the short-rate is modeled as a single stochastic factor:

$$dr(t) = K(\theta - r(t))dt + \sigma dW(t)$$

Here, K, θ, and σ are constants, and σ is the instantaneous standard deviation. $W(t)$ is the random Wiener process. The Vasicek follows an Ornstein-Uhlenbeck process, where the model reverts around the mean, θ, with K, the speed of mean reversion. As a result, the interest rates may become negative, which is an undesirable property in most normal economic conditions.

To help us understand this model, the following code generates a list of interest rates:

```
In [ ]:
    import math
    import numpy as np

    def vasicek(r0, K, theta, sigma, T=1., N=10, seed=777):
        np.random.seed(seed)
        dt = T/float(N)
        rates = [r0]
        for i in range(N):
            dr = K*(theta-rates[-1])*dt + \
                sigma*math.sqrt(dt)*np.random.normal()
            rates.append(rates[-1]+dr)

        return range(N+1), rates
```

The `vasicek()` function returns a list of time periods and interest rates from the Vasicek model. It takes in a number of input arguments: `r0` is the initial rate of interest at *t=0*; `K`, `theta`, and `sigma` are constants; `T` is the period in terms of number of years; `N` is the number of intervals for the modeling process; and `seed` is the initialization value for NumPy's standard normal random-number generator.

Assume that the current interest rate is close to zero at 0.5%, the long term mean level `theta` is `0.15`, and the instantaneous volatility `sigma` is 5%. We will use a `T` value of `10` and an `N` value of `200` to model the interest rates for different speeds of mean reversion, `K`, with values of `0.002`, `0.02`, and `0.2`:

```
In [ ]:
    %pylab inline

    fig = plt.figure(figsize=(12, 8))
```

```
for K in [0.002, 0.02, 0.2]:
    x, y = vasicek(0.005, K, 0.15, 0.05, T=10, N=200)
    plot(x,y, label='K=%s'%K)
    pylab.legend(loc='upper left');

pylab.legend(loc='upper left')
pylab.xlabel('Vasicek model');
```

After running the preceding commands, we get the following graph:

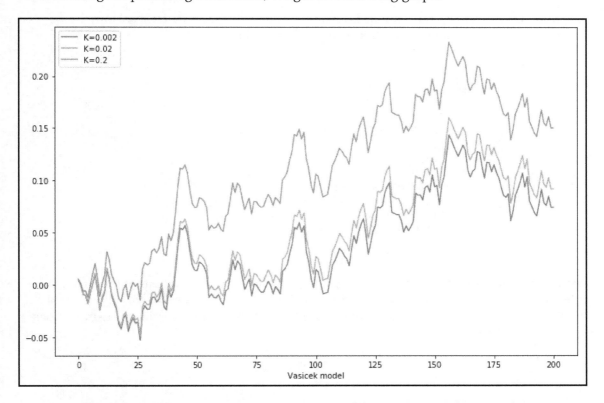

In this example, we are running just one simulation to see what the interest rates from the Vasicek model looks like. Observe that interest rates did become negative at some point. With higher levels of speed for mean reversion, K, the process reaches its long-term level of 0.15 sooner.

The Cox-Ingersoll-Ross model

The **Cox-Ingersoll-Ross** (**CIR**) model is a one-factor model that was proposed to address the negative interest rates found in the Vasicek model. The process is given as follows:

$$dr(t) = K(\theta - r(t))dt + \sigma\sqrt{r(t)}dW(t)$$

The term $\sqrt{r(t)}$ increases the standard deviation as the short-rate increases. Now the `vasicek()` function can be rewritten as the CIR model in Python:

```
In [ ]:
    import math
    import numpy as np

    def CIR(r0, K, theta, sigma, T=1.,N=10,seed=777):
        np.random.seed(seed)
        dt = T/float(N)
        rates = [r0]
        for i in range(N):
            dr = K*(theta-rates[-1])*dt + \
                sigma*math.sqrt(rates[-1])*\
                math.sqrt(dt)*np.random.normal()
            rates.append(rates[-1] + dr)

        return range(N+1), rates
```

Using the same example given in the *The Vasicek model* section, assume that the current interest rate is 0.5%, `theta` is `0.15`, and `sigma` is `0.05`. We will use a `T` value of `10` and `N` of `200` to model the interest rates with different speeds of mean reversion, `K`, using values of `0.002, 0.02`, and `0.2`:

```
In [ ] :
    %pylab inline

    fig = plt.figure(figsize=(12, 8))

    for K in [0.002, 0.02, 0.2]:
        x, y = CIR(0.005, K, 0.15, 0.05, T=10, N=200)
        plot(x,y, label='K=%s'%K)

    pylab.legend(loc='upper left')
    pylab.xlabel('CRR model');
```

Here is the output of the preceding commands:

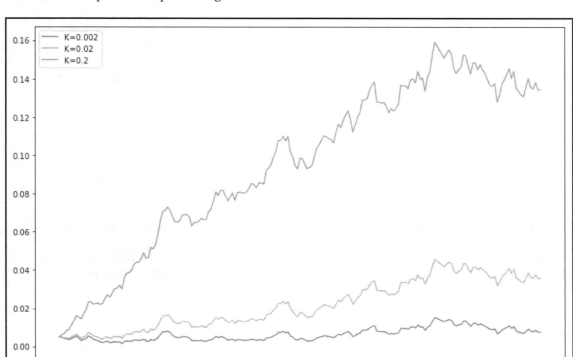

Observe that the CIR interest model does not have negative interest rate values.

The Rendleman and Bartter model

In the Rendleman and Bartter model, the short-rate process is given as follows:

$$dr(t) = \theta r(t)dt + \sigma r(t)dW(t)$$

Here, the instantaneous drift is $\theta r(t)$ with an instantaneous standard deviation, $\sigma r(t)$. The Rendleman and Bartter model can be thought of as a geometric Brownian motion, akin to a stock price stochastic process that is log-normally distributed. This model lacks the property of mean reversion. Mean reversion is a phenomenon where the interest rates seem to be pulled back toward a long-term average level.

The following Python code models the Rendleman and Bartter interest-rate process:

```
In [ ]:
    import math
    import numpy as np

    def rendleman_bartter(r0, theta, sigma, T=1.,N=10,seed=777):
        np.random.seed(seed)
        dt = T/float(N)
        rates = [r0]
        for i in range(N):
            dr = theta*rates[-1]*dt + \
                sigma*rates[-1]*math.sqrt(dt)*np.random.normal()
            rates.append(rates[-1] + dr)

        return range(N+1), rates
```

We will continue to use the example from the previous sections and compare the model.

Assume that the current interest rate is 0.5%, and `sigma` is `0.05`. We will use a `T` value of `10` and `N` of `200` to model the interest rates with different instantaneous drift, `theta`, using values of `0.01`, `0.05`, and `0.1`:

```
In [ ]:
    %pylab inline

    fig = plt.figure(figsize=(12, 8))

    for theta in [0.01, 0.05, 0.1]:
        x, y = rendleman_bartter(0.005, theta, 0.05, T=10, N=200)
        plot(x,y, label='theta=%s'%theta)

    pylab.legend(loc='upper left')
    pylab.xlabel('Rendleman and Bartter model');
```

The following graph is the output for the preceding commands:

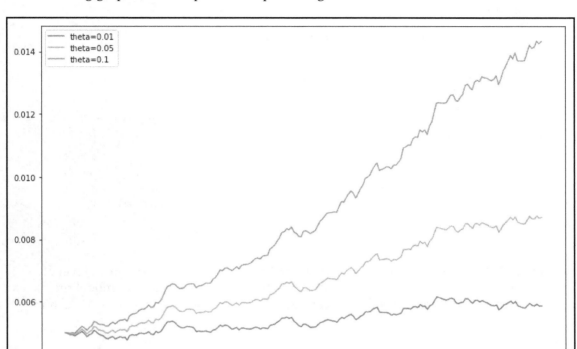

In general, this model lacks the property of mean reversion and grows toward a long-term average level.

The Brennan and Schwartz model

The Brennan and Schwartz model is a two-factor model where the short-rate reverts toward a long-term rate as the mean, which also follows a stochastic process. The short-rate process is given as follows:

$$dr(t) = K(\theta - t(t))dt + \sigma r(t)dW(t)$$

It can be seen that the Brennan and Schwartz model is another form of a geometric Brownian motion.

Our Python code can now be implemented as follows:

```
In [ ]:
    import math
    import numpy as np

    def brennan_schwartz(r0, K, theta, sigma, T=1., N=10, seed=777):
        np.random.seed(seed)
        dt = T/float(N)
        rates = [r0]
        for i in range(N):
            dr = K*(theta-rates[-1])*dt + \
                sigma*rates[-1]*math.sqrt(dt)*np.random.normal()
            rates.append(rates[-1] + dr)

        return range(N+1), rates
```

Assume that the current interest rate remains at 0.5%, and the long-term mean level of theta is 0.006. sigma is 0.05. We will use a T value of 10 and N of 200 to model the interest rates with different speeds of mean reversion, K, using values of 0.2, 0.02, and 0.002:

```
In [ ]:
    %pylab inline

    fig = plt.figure(figsize=(12, 8))

    for K in [0.2, 0.02, 0.002]:
        x, y = brennan_schwartz(0.005, K, 0.006, 0.05, T=10, N=200)
        plot(x,y, label='K=%s'%K)

    pylab.legend(loc='upper left')
    pylab.xlabel('Brennan and Schwartz model');
```

After running the preceding commands, we will get the following output:

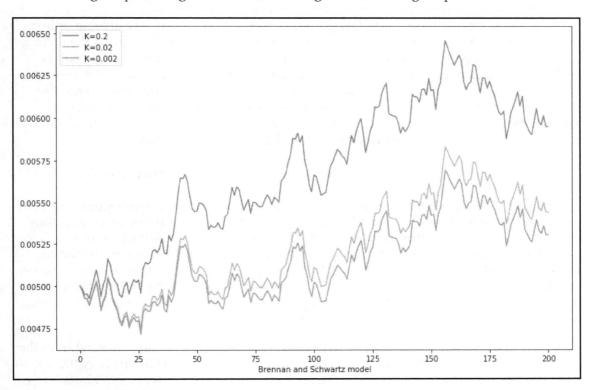

When k is 0.2, the speed of the mean reversion is fastest toward the long-term mean of 0.006.

Bond options

When bond issuers, such as corporations, issue bonds, one of the risks they face is the interest rate risk. When interest rates decrease, bond prices increase. While existing bondholders will find their bonds more valuable, bond issuers, on the other hand, find themselves in a losing position, since they will be issuing higher-interest payments than the prevailing interest rate. Conversely, when interest rates increase, bond issuers are at an advantage, since they are able to continue issuing the same low-interest payments as agreed to on the bond-contract specifications.

To capitalize on interest-rate changes, bond issuers may embed options within a bond. This allows the issuer the right, but not the obligation, to buy or sell the issued bond at a predetermined price during a specified period of time. An American type of bond option allows the issuer to exercise the rights of the option at any point in time during the lifetime of a bond. A European type of bond option allows the issuer to exercise the rights of the option on a specific date. The exact style of the date of exercise varies from bond option to bond option. Some issuers may choose to exercise the right of the bond option when the bond has been in circulation in the market for over a year. Some issuers may choose to exercise the bond option on one of several specific dates. Regardless of the exercise dates of the bond, you may price the bond with an embedded option, as follows:

Price of bond = price of bond with no option - price of embedded option

The pricing of a bond with no option is fairly straightforward: the present value of the bond to be received at a future date, including all coupon payments. A number of assumptions are to be made about the theoretical interest rates in the future at which the coupon payments may be reinvested. One such assumption might be the movement of interest rates as implied by short-rate models, which we covered in the preceding section, *Short-rate modeling*. Another assumption might be the movement of interest rates within a binomial or trinomial tree. For simplicity, in bond-pricing studies, we will price zero-coupon bonds that will not issue coupons during the lifetime of the bond.

To price an option, one would have to determine available exercise dates. Starting from the future value of the bond, the bond price is compared against the exercise price of the option and traverses back to the present time using a numerical procedure, such as a binomial tree. This price comparison is performed at time points when the bond option may be exercised. With the no-arbitrage theory, accounting for the present excess values of the bond when exercised, we obtain the price of the option. For simplicity, in bond-pricing studies in the later section of this chapter, *Pricing callable bond options*, we will treat the embedded option of zero-coupon bonds as an American option.

Callable bonds

In an economic condition where there are high interest rates, bond issuers are likely at risk of facing an interest-rate decrease and having to continue to issue higher-interest payments than the prevailing interest rate. As such, they may choose to issue callable bonds. A callable bond contains an embedded agreement to redeem the bond at agreed dates. Existing bond holders are considered to have sold a call option to the bond issuer.

In the event that interest rates fall and the corporation has the right to exercise the option to buy back the bond during that period at a specific price, they may choose to do so. The company can then issue new bonds at lower interest rates. This also means that the company is able to raise more capital in the form of higher bond prices.

Puttable bonds

Unlike callable bonds, the owner of puttable bonds has the right, but not the obligation, to sell the bond back to the issuer at an agreed price during a certain period. Owners of puttable bonds are considered to have bought a put option from the bond issuer. When interest rates increase, values of existing bonds become less valuable and puttable bond holders are more incentivized to exercise the right to sell the bond at a higher exercise price. Since puttable bonds are more beneficial to buyers than to the issuers, they are generally less common than callable bonds. Variants of puttable bonds can be found in the form of loan and deposit instruments. A customer who has placed a fixed-rate deposit with a financial institution receives interest payments on specified dates. They are entitled to withdraw the deposit at any time. As such, a fixed-rate deposit instrument can be thought of as a bond with an embedded American put option.

An investor who wishes to borrow money from a bank enters a loan agreement, making interest payments during the lifetime of the agreement until the debt, together with the principal amount and agreed interest, is fully repaid. The bank can be considered as buying a put option on a bond. Under certain circumstances, the bank may exercise the right to redeem the full value of the loan agreement.

Thus, the price of puttable bonds can be thought of as follows:

Price of puttable bond = price of bond with no option + price of put option

Convertible bonds

Convertible bonds are issued by companies and contain an embedded option that allows the holder to convert the bond into a number of shares of common stock. The amount of shares to be converted for a bond is defined as the conversion ratio, which is determined such that the dollar amount of shares is the same as the value of the bond.

Convertible bonds have similarities with callable bonds. They allow the bond holders to exercise the bond for an equivalent amount of shares at the specified conversion ratio at agreed times. Convertible bonds typically issue lower coupon rates than non-convertible bonds, to compensate for the additional value of the right to exercise.

When convertible bond holders exercise their rights into stocks, the company's debts are reduced. On the other hand, the company's stocks become more diluted as the number of shares in the circulation increases, and the company's stock price is expected to fall.

As the company's stock price increases, convertible bond prices tend to increase. Conversely, as the company's stock price decreases, convertible bond prices tend to decrease.

Preferred stocks

Preferred stocks are stocks that have bond-like qualities. Owners of preferred stocks have seniority of claim on dividend payments over common stocks, which are usually negotiated as a fixed percentage of their par value. Although there is no guarantee of dividend payments, all dividends are paid on preferred stock first over common stock. In certain agreements on preferred stocks, dividends that are not paid as agreed may accumulate until they are paid at a later time. These preferred stocks are known as **cumulative**.

Prices of preferred stocks typically move in tandem with their common stock. They may have voting rights associated with common shareholders. In the event of bankruptcy, preferred stocks have a first lien of its par value upon liquidation.

Pricing a callable bond option

In this section, we will take a look at pricing a callable bond. We assume that the bond to be priced is a zero-coupon paying bond with an embedded European call option. The price of a callable bond can be thought of as follows:

Price of callable bond = price of bond with no option – price of call option

Pricing a zero-coupon bond by the Vasicek model

The value of a zero-coupon bond with a par value of 1 at time t and a prevailing interest rate, r, is defined as follows:

$$P(t) = e^{-rdt}$$

Since the interest rate, *r*, is always changing, we rewrite the zero-coupon bond as follows:

$$P(t) = e^{-\int_t^T r(s)dS}$$

Now, the interest rate, *r*, is a stochastic process that accounts for the price of the bond from time *t* to *T*, where *T* is the time to maturity of the zero-coupon bond.

To model the interest rate, *r*, we can use one of the short-rate models as a stochastic process. For this purpose, we will use the Vasicek model to model the short-rate process.

The expectation of a log-normally distributed variable, *X*, is given as follows:

$$X = e^u$$

$$E[X] = E[e^u] = e^{u + \frac{\sigma^2}{2}}$$

Taking moments of the log-normally distributed variable *X*:

$$E[e^{su}] = e^{su + \frac{s^2 \sigma^2}{2}}$$

We obtained the expected value of a log-normally distributed variable, which we will use in the interest-rate process for the zero-coupon bond.

Remember the Vasicek short-rate process model:

$$dr(t) = K(\theta - r(t))dt + \sigma dW(t)$$

Then, *r(t)* is derived as follows:

$$r(t) = \theta + (r_0 - \theta)e^{-kt} + \sigma e^{-kt} \int_0^t e^{ks} dB$$

Using the characteristic equation and the interest-rate movements of the Vasicek model, we can rewrite the zero-coupon bond price in terms of expectations:

$$P(t) = E\left[e^{-\int_t^l r(s)ds}\right]$$

$$P(\tau) = A(\tau)e^{-r_t B(\tau)}$$

Here:

$$A(\tau) = e^{\left(\theta - \frac{\sigma^2}{2k^2}\right)(B(\tau) - \tau) - \frac{\sigma^2}{4k}B(\tau)^2}$$

$$B(\tau) = \frac{1 - e^{-k\tau}}{k}$$

$$\tau = T - t$$

The Python implementation of the zero-coupon bond price is given in the `exact_zcb` function:

```
In [ ]:
    import numpy as np
    import math

    def exact_zcb(theta, kappa, sigma, tau, r0=0.):
        B = (1 - np.exp(-kappa*tau)) / kappa
        A = np.exp((theta-(sigma**2)/(2*(kappa**2)))*(B-tau) - \
                   (sigma**2)/(4*kappa)*(B**2))
        return A * np.exp(-r0*B)
```

For example, we are interested in finding out the prices of zero-coupon bond prices for a number of maturities. We model the Vasicek short-rate process with a `theta` value of `0.5`, `kappa` value of `0.02`, `sigma` value of `0.03`, and an initial interest rate, `r0`, of `0.015`.

Plugging these values into the `exact_zcb` function, we obtain zero-coupon bond prices, for the time period from 0 to 25 years with intervals of 0.5 years, and plot out the graph:

```
In [ ]:
    Ts = np.r_[0.0:25.5:0.5]
    zcbs = [exact_zcb(0.5, 0.02, 0.03, t, 0.015) for t in Ts]
In [ ]:
    %pylab inline

    fig = plt.figure(figsize=(12, 8))
    plt.title("Zero Coupon Bond (ZCB) Values by Time")
    plt.plot(Ts, zcbs, label='ZCB')
    plt.ylabel("Value ($)")
    plt.xlabel("Time in years")
    plt.legend()
    plt.grid(True)
    plt.show()
```

The following graph is the output for the preceding commands:

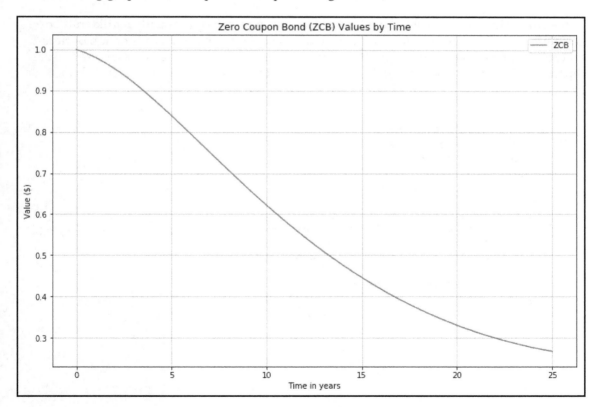

The value of early exercise

Issuers of callable bonds may redeem the bond at an agreed price, as specified in the contract. To price such a bond, the discounted early-exercise values can be defined as follows:

$$\text{discounted early exercise value} = ke^{-rt}$$

Here, k is the price ratio of the strike price to the par value and r is the interest rate for the strike price.

The Python implementation of the early-exercise option can then be written as follows:

```
In [ ]:
    import math

    def exercise_value(K, R, t):
        return K*math.exp(-R*t)
```

In the preceding example, we are interested in valuing a call option with a strike ratio of 0.95 and an initial interest rate of 1.5%. We can then plot the values as a function of time and superimpose them onto a graph of zero-coupon bond prices to give us a better visual representation of the relationship between zero-coupon bond prices and callable bond prices:

```
In [ ]:
    Ts = np.r_[0.0:25.5:0.5]
    Ks = [exercise_value(0.95, 0.015, t) for t in Ts]
    zcbs = [exact_zcb(0.5, 0.02, 0.03, t, 0.015) for t in Ts]
In [ ]:
    import matplotlib.pyplot as plt

    fig = plt.figure(figsize=(12, 8))
    plt.title("Zero Coupon Bond (ZCB) and Strike (K) Values by Time")
    plt.plot(Ts, zcbs, label='ZCB')
    plt.plot(Ts, Ks, label='K', linestyle="--", marker=".")
    plt.ylabel("Value ($)")
    plt.xlabel("Time in years")
    plt.legend()
    plt.grid(True)
    plt.show()
```

Here is the output for the preceding commands:

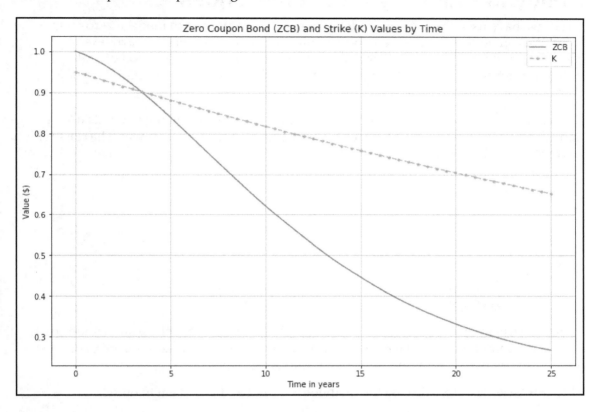

From the preceding graph, we can approximate the price of callable zero-coupon bond prices. Since the bond issuer owns the call, the price of the callable zero-coupon bond can be stated as follows:

$$\text{callable zero-coupon bond price} = min(ZCB, K)$$

This callable bond price is an approximation, given the current interest-rate level. The next step would be to treat early-exercise by going through a form of policy iteration, which is a cycle used to determine optimum early-exercise values and their effect on other nodes, and check whether they become due for an early exercise. In practice, such an iteration only occurs once.

Policy iteration by finite differences

So far, we have used the Vasicek model in our short-rate process to model a zero-coupon bond. We can undergo policy iteration by finite differences to check for early-exercise conditions and their effect on other nodes. We will use the implicit method of finite differences for the numerical pricing procedure, as discussed in Chapter 4, *Numerical Procedures for Pricing Options*.

Let's create a class named VasicekCZCB, which will incorporate all the methods used to implement the pricing of callable zero-coupon bonds by the Vasicek model. The class with its constructor definition is given as follows:

```
import math
import numpy as np
import scipy.stats as st


class VasicekCZCB:

    def __init__(self):
        self.norminv = st.distributions.norm.ppf
        self.norm = st.distributions.norm.cdf
```

In the constructor, the norminv and normv variable is made available to all methods that require the computation of the inverse normal cumulative distribution function and the normal cumulative distribution function of SciPy, respectively.

With this base class, let's discuss the methods required and add them to our class:

- Add the vasicek_czcb_values() method as the point of entry to begin the pricing process. The r0 variable is the short-rate at time *t=0*; R is the strike zero rate for the bond price; ratio is the strike price per par value of the bond; T is the time to maturity; sigma is the volatility of the short-rate, r; kappa is the rate of mean reversion; theta is the mean of the short-rate process; M is the number of steps in the finite differences scheme; prob is the probability on the normal distribution curve used by the vasicek_limits method later to determine short-rates; max_policy_iter is the maximum number of policy iterations used to find early-exercise nodes; grid_struct_const is the maximum threshold of dt movement that determines N in the calculate_N() method; and rs is the list of interest rates from which the short-rate process follows.

This method returns a list of evenly-spaced short-rates and a list of option prices, and is written as follows:

```
def vasicek_czcb_values(self, r0, R, ratio, T, sigma, kappa, theta,
                        M, prob=1e-6, max_policy_iter=10,
                        grid_struct_const=0.25, rs=None):
    (r_min, dr, N, dtau) = \
        self.vasicek_params(r0, M, sigma, kappa, theta,
                            T, prob, grid_struct_const, rs)
    r = np.r_[0:N]*dr + r_min
    v_mplus1 = np.ones(N)

    for i in range(1, M+1):
        K = self.exercise_call_price(R, ratio, i*dtau)
        eex = np.ones(N)*K
        (subdiagonal, diagonal, superdiagonal) = \
            self.vasicek_diagonals(
                sigma, kappa, theta, r_min, dr, N, dtau)
        (v_mplus1, iterations) = \
            self.iterate(subdiagonal, diagonal, superdiagonal,
                         v_mplus1, eex, max_policy_iter)
    return r, v_mplus1
```

- Add the `vasicek_params()` method to compute the implicit scheme parameters for the Vasicek model. It returns a tuple of `r_min`, `dr`, `N`, and `dt`. If no value is supplied to `rs`, values of `r_min` to `r_max` will be automatically generated by the `vasicek_limits()` method as a function of `prob` following a normal distribution. This method is written as follows:

```
def vasicek_params(self, r0, M, sigma, kappa, theta, T,
                   prob, grid_struct_const=0.25, rs=None):
    if rs is not None:
        (r_min, r_max) = (rs[0], rs[-1])
    else:
        (r_min, r_max) = self.vasicek_limits(
            r0, sigma, kappa, theta, T, prob)

    dt = T/float(M)
    N = self.calculate_N(grid_struct_const, dt, sigma, r_max,
r_min)
    dr = (r_max-r_min)/(N-1)

    return (r_min, dr, N, dt)
```

- Add the `calculate_N()` method, which is used by the `vasicek_params()` method to compute the grid size parameter, `N`. This method is written as follows:

```
def calculate_N(self, max_structure_const, dt, sigma, r_max,
r_min):
    N = 0
    while True:
        N += 1
        grid_structure_interval = \
            dt*(sigma**2)/(((r_max-r_min)/float(N))**2)
        if grid_structure_interval > max_structure_const:
            break
    return N
```

- Add the `vasicek_limits()` method to compute the minimum and maximum of the Vasicek interest rate process by a normal distribution process. The expected value of the short-rate process, `r(t)`, under the Vasicek model is given as follows:

$$E[r(t)] = \theta + (r_0 - \theta)e^{-kt}$$

The variance is defined as follows:

$$Var[r(t)] = \frac{\sigma^2}{2k}(1 - e^{-2kt})$$

This method returns a tuple of the minimum and maximum interest-rate level as defined by the probability for the normal distribution process, and is written as follows:

```
def vasicek_limits(self, r0, sigma, kappa, theta, T, prob=1e-6):
    er = theta+(r0-theta)*math.exp(-kappa*T)
    variance = (sigma**2)*T if kappa==0 else \
                (sigma**2)/(2*kappa)*(1-math.exp(-2*kappa*T))
    stdev = math.sqrt(variance)
    r_min = self.norminv(prob, er, stdev)
    r_max = self.norminv(1-prob, er, stdev)
    return (r_min, r_max)
```

- Add the `vasicek_diagonals()` method, which returns the diagonals of the implicit scheme of finite differences, where:

$$sub - diagonals, a = k(\theta - r_i)\frac{dt}{2dr} - \frac{1}{2}\sigma^2\frac{dt}{dr^2}$$

$$diagonals, b = 1 + r_i dt + \sigma^2\frac{dt}{dr^2}$$

$$super - diagonals, c = k(\theta - r_i)\frac{dt}{2dr} - \frac{1}{2}\sigma^2\frac{dt}{dr^2}$$

The boundary conditions are implemented using Neumann boundary conditions. This method is written as follows:

```python
def vasicek_diagonals(self, sigma, kappa, theta, r_min,
                      dr, N, dtau):
    rn = np.r_[0:N]*dr + r_min
    subdiagonals = kappa*(theta-rn)*dtau/(2*dr) - \
                   0.5*(sigma**2)*dtau/(dr**2)
    diagonals = 1 + rn*dtau + sigma**2*dtau/(dr**2)
    superdiagonals = -kappa*(theta-rn)*dtau/(2*dr) - \
                     0.5*(sigma**2)*dtau/(dr**2)

    # Implement boundary conditions.
    if N > 0:
        v_subd0 = subdiagonals[0]
        superdiagonals[0] = superdiagonals[0]-subdiagonals[0]
        diagonals[0] += 2*v_subd0
        subdiagonals[0] = 0

    if N > 1:
        v_superd_last = superdiagonals[-1]
        superdiagonals[-1] = superdiagonals[-1] - subdiagonals[-1]
        diagonals[-1] += 2*v_superd_last
        superdiagonals[-1] = 0

    return (subdiagonals, diagonals, superdiagonals)
```

The Neumann boundary condition specifies the boundaries of a given ordinary or partial differential equation. Further information can be found at http://mathworld.wolfram.com/NeumannBoundaryConditions.html.

- Add the `check_exercise()` method, which returns a list of Boolean values, indicating the indices suggesting optimum payoff from an early exercise. This method is written as follows:

```
def check_exercise(self, V, eex):
    return V > eex
```

- Add the `exercise_call_price()` method, which returns the discounted value of the strike price as a ratio, written as follows:

```
def exercise_call_price(self, R, ratio, tau):
    K = ratio*np.exp(-R*tau)
    return K
```

- Add the `vasicek_policy_diagonals()` method, which is being called by the policy-iteration procedure that updates the sub-diagonals, diagonals, and super-diagonals for one iteration. In indices, where an early exercise is carried out, the sub-diagonals and super-diagonals will have these values set to 0 and the remaining values on the diagonals. The method returns comma-separated values of the new sub-diagonal, diagonal, and super-diagonal values. This method is written as follows:

```
def vasicek_policy_diagonals(self, subdiagonal, diagonal, \
                             superdiagonal, v_old, v_new, eex):
    has_early_exercise = self.check_exercise(v_new, eex)
    subdiagonal[has_early_exercise] = 0
    superdiagonal[has_early_exercise] = 0
    policy = v_old/eex
    policy_values = policy[has_early_exercise]
    diagonal[has_early_exercise] = policy_values
    return (subdiagonal, diagonal, superdiagonal)
```

- Add the `iterate()` method, which implements the implicit scheme of finite differences by performing a policy iteration, where each cycle involves solving the tridiagonal systems of equations, calling the `vasicek_policy_diagonals()` method to update the three diagonals, and returns the callable zero-coupon bond price if there are no further early-exercise opportunities. It also returns the number of policy iterations performed. This method is written as follows:

```
def iterate(self, subdiagonal, diagonal, superdiagonal,
            v_old, eex, max_policy_iter=10):
    v_mplus1 = v_old
```

```
v_m = v_old
change = np.zeros(len(v_old))
prev_changes = np.zeros(len(v_old))

iterations = 0
while iterations <= max_policy_iter:
    iterations += 1

    v_mplus1 = self.tridiagonal_solve(
            subdiagonal, diagonal, superdiagonal, v_old)
    subdiagonal, diagonal, superdiagonal = \
        self.vasicek_policy_diagonals(
            subdiagonal, diagonal, superdiagonal,
            v_old, v_mplus1, eex)

    is_eex = self.check_exercise(v_mplus1, eex)
    change[is_eex] = 1

    if iterations > 1:
        change[v_mplus1 != v_m] = 1

    is_no_more_eex = False if True in is_eex else True
    if is_no_more_eex:
        break

    v_mplus1[is_eex] = eex[is_eex]
    changes = (change == prev_changes)

    is_no_further_changes = all((x == 1) for x in changes)
    if is_no_further_changes:
        break

    prev_changes = change
    v_m = v_mplus1

return v_mplus1, iterations-1
```

- Add the `tridiagonal_solve()` method, which implements the Thomas algorithm to solve tridiagonal systems of equations. The systems of equations may be written as follows:

$$a_i x_{i-1} + b_i x_i + c_i a_i x_{i+1} = d_i$$

This equation is represented in matrix form:

$$\begin{bmatrix} b_1 & c_1 & 0 & 0 \\ a_2 & b_2 & \ddots & 0 \\ 0 & \ddots & \ddots & c_{n-1} \\ 0 & 0 & a_n & b_n \end{bmatrix} \begin{bmatrix} x_1 \\ x_1 \\ \vdots \\ x_n \end{bmatrix} = \begin{bmatrix} d_1 \\ d_1 \\ \vdots \\ d_n \end{bmatrix}$$

Here, a is a list for the sub-diagonals, b is a list for the diagonal, and c is the super-diagonal of the matrix.

 The Thomas algorithm is a matrix algorithm for solving tridiagonal systems of equations using a simplified form of Gaussian elimination. Further information can be found at http://faculty.washington.edu/ finlayso/ebook/algebraic/advanced/LUtri.htm.

The tridiagonal_solve() method is written as follows:

```
def tridiagonal_solve(self, a, b, c, d):
    nf = len(a)  # Number of equations
    ac, bc, cc, dc = map(np.array, (a, b, c, d))  # Copy the array
    for it in range(1, nf):
        mc = ac[it]/bc[it-1]
        bc[it] = bc[it] - mc*cc[it-1]
        dc[it] = dc[it] - mc*dc[it-1]

    xc = ac
    xc[-1] = dc[-1]/bc[-1]

    for il in range(nf-2, -1, -1):
        xc[il] = (dc[il]-cc[il]*xc[il+1])/bc[il]

    del bc, cc, dc  # Delete variables from memory

    return xc
```

With these methods defined, we can now run our code and price a callable zero-coupon bond with the Vasicek model.

Assume that we run this model with the following parameters: r0 is 0.05, R is 0.05, ratio is 0.95, sigma is 0.03, kappa is 0.15, theta is 0.05, prob is 1e-6, M is 250, max_policy_iter is 10, grid_struc_interval is 0.25, and we are interested in the values of the interest rates between 0% and 2%.

The following Python code demonstrates this model for maturities of 1 year, 5 years, 7 years, 10 years, and 20 years:

```
In [ ]:
    r0 = 0.05
    R = 0.05
    ratio = 0.95
    sigma = 0.03
    kappa = 0.15
    theta = 0.05
    prob = 1e-6
    M = 250
    max_policy_iter=10
    grid_struct_interval = 0.25
    rs = np.r_[0.0:2.0:0.1]
In [ ]:
    vasicek = VasicekCZCB()
    r, vals = vasicek.vasicek_czcb_values(
        r0, R, ratio, 1., sigma, kappa, theta,
        M, prob, max_policy_iter, grid_struct_interval, rs)
In [ ]:
    %pylab inline

    fig = plt.figure(figsize=(12, 8))
    plt.title("Callable Zero Coupon Bond Values by r")
    plt.plot(r, vals, label='1 yr')

    for T in [5., 7., 10., 20.]:
        r, vals = vasicek.vasicek_czcb_values(
            r0, R, ratio, T, sigma, kappa, theta,
            M, prob, max_policy_iter, grid_struct_interval, rs)
        plt.plot(r, vals, label=str(T)+' yr', linestyle="--", marker=".")

    plt.ylabel("Value ($)")
    plt.xlabel("r")
    plt.legend()
    plt.grid(True)
    plt.show()
```

After running the preceding commands, you should get the following output:

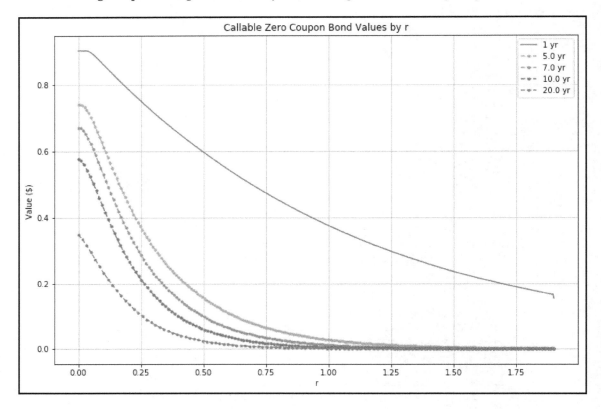

We obtained the theoretical values of pricing callable zero-coupon bonds for various maturities for various interest rates.

Other considerations in callable bond pricing

In pricing callable zero-coupon bonds, we used the Vasicek interest-rate process to model interest-rate movement with the aid of a normal distribution process. In *The Vasicek model* section, we demonstrated that the Vasicek model can produce negative interest rates, which may not be practical for most economic cycles. Quantitative analysts often use more than one model in derivative pricing to obtain realistic results. The CIR and Hull-White models are some of the commonly-discussed models in financial studies. The limitation on these models is that they involve only one factor, or a single source of uncertainty.

We also looked at the implicit scheme of finite differences for the policy iteration of the early exercise. Another method of consideration is the Crank-Nicolson method of finite differences. Other methods include the Monte Carlo simulation for calibration of this model.

Finally, we obtained a final list of short-rates and callable bond prices. To infer a fair value for the callable bond for a particular short-rate, interpolation of the list of bond prices is required. Often, the linear interpolation method is used. Other interpolation methods of consideration are the cubic and spline interpolation methods.

Summary

In this chapter, we focused on interest-rate and related derivative pricing with Python. Most bonds, such as US Treasury bonds, pay a fixed amount of interest semi-annually, while other bonds may pay quarterly or annually. It is a characteristic of bonds that their prices are closely related to current interest-rate levels in an inverse manner. The normal or positive yield curve, where long-term interest rates are higher than short-term interest rates, is said to be upward sloping. In certain economic conditions, the yield curve can be inverted and is said to be downward sloping.

A zero-coupon bond is a bond that pays no coupons during its lifetime, except upon maturity when the principal or face value is repaid. We implemented a simple zero-coupon bond calculator in Python.

The yield curve can be derived from the short-term zero or spot rates of securities, such as zero-coupon bonds, T-bills, notes, and eurodollar deposits using a bootstrapping process. With Python, we used a lot of bond information to plot a yield curve, and derived forward rates, yield-to-maturity, and bond prices from the yield curve.

Two important metrics to bond traders are duration and convexity. Duration is the sensitivity measure of bond prices to yield changes. Convexity is the sensitivity measure of the duration of a bond to yield changes. We implemented calculations using the modified duration model and convexity calculator in Python.

Short-rate models are frequently used in the evaluation of interest-rate derivatives. Interest-rate modeling is a fairly complex topic, since they are affected by a multitude of factors, such as economic states, political decisions, government intervention, and the laws of supply and demand. A number of interest-rate models have been proposed to account for the various characteristics of interest rates. Some of the interest-rate models we discussed include the Vasicek, CIR, and Rendleman and Bartter models.

Bond issuers may embed options within a bond to allow them the right, but not the obligation, to buy or sell the issued bond at a predetermined price during a specified period of time. The price of a callable bond can be thought of as the price difference of a bond without an option and the price of the embedded call option. Using Python, we took a look at pricing a callable zero-coupon bond by applying the Vasicek model to the implicit method of finite differences. This method is, however, just one of the many methods that quantitative analysts use in bond-options modeling.

In the next chapter, we will discuss the statistical analysis of time-series data.

Statistical Analysis of Time Series Data

6

In financial portfolios, the returns on their constituent assets depend on a number of factors, such as macroeconomic and microeconomical conditions, and various financial variables. As the number of factors increases, so does the complexity involved in modeling portfolio behavior. Given that computing resources are finite, coupled with time constraints, performing an extra computation for a new factor only increases the bottleneck on portfolio modeling calculations. A linear technique for dimensionality reduction is **Principal Component Analysis (PCA)**. As its name suggests, PCA breaks down the movement of portfolio asset prices into its principal components, or common factors, for further statistical analysis. Common factors that don't explain much of the movement of the portfolio assets receive less weighting in their factors and are usually ignored. By keeping the most useful factors, portfolio analysis can be greatly simplified without compromising on computational time and space costs.

In statistical analysis of time series data, it is important for the data to be stationary in order to avoid spurious regression. Non-stationary data may be generated by an underlying process that is affected by a trend, a seasonal effect, presence of a unit root, or a combination of all three. The statistical properties of non-stationary data, such as mean and variance, changes over time. Non-stationary data needs to be transformed into stationary data for statistical analysis to produce consistent and reliable results. This can be achieved by removing the trend and seasonality components. Stationary data can thereafter be used for prediction or forecasting.

In this chapter, we will cover the following topics:

- Performing PCA on the Dow and its 30 components
- Reconstructing the Dow index
- Understanding the difference between stationary and non-stationary data
- Checking data for stationarity
- Types of stationary and non-stationary processes

- Using the Augmented Dickey-Fuller Test to test the presence of a unit root
- Making stationary data by detrending, differencing, and seasonal decomposing
- Using an Autoregressive Integrated Moving Average for time series prediction and forecasting

The Dow Jones industrial average and its 30 components

The **Dow Jones Industrial Average** (DJIA) is a stock market index that comprises the 30 largest US companies. Commonly known as the **Dow**, it is owned by S&P Dow Jones Indices LLC and computed on a price-weighted basis (see `https://us.spindices.com/index-family/us-equity/dow-jones-averages` for more information on the Dow).

This section involves downloading the datasets of Dow and its components into `pandas` DataFrame objects for use in later sections of this chapter.

Downloading Dow component datasets from Quandl

The following code retrieves the Dow component datasets from Quandl. The data provider that we will be using is WIKI Prices, a community formed by members of the public and that provides datasets free of charge back to the public. Such data isn't free from errors, so please use them with caution. At the time of writing, this data feed is no longer supported actively by the Quandl community, though past datasets are still available for use. We will download historical daily closing prices for 2017:

```
In [ ]:
    import quandl

    QUANDL_API_KEY = 'BCzkk3NDWt7H9yjzx-DY'  # Your own Quandl key here
    quandl.ApiConfig.api_key = QUANDL_API_KEY

    SYMBOLS = [
        'AAPL','MMM', 'AXP', 'BA', 'CAT',
        'CVX', 'CSCO', 'KO', 'DD', 'XOM',
        'GS', 'HD', 'IBM', 'INTC', 'JNJ',
        'JPM', 'MCD', 'MRK', 'MSFT', 'NKE',
        'PFE', 'PG', 'UNH', 'UTX', 'TRV',
        'VZ', 'V', 'WMT', 'WBA', 'DIS',
```

```
        ]

        wiki_symbols = ['WIKI/%s'%symbol for symbol in SYMBOLS]
        df_components = quandl.get(
            wiki_symbols,
            start_date='2017-01-01',
            end_date='2017-12-31',
            column_index=11)
        df_components.columns = SYMBOLS  # Renaming the columns
```

The `wiki_symbols` variable contains a list of Quandl codes that we use for downloading. Notice that in the parameter arguments of `quandl.get()`, we specified `column_index=11`. This tells Quandl to download only the 11th column of each dataset, which coincides with the adjusted daily closing prices. The datasets are downloaded into our `df_components` variable as a single `pandas` DataFrame object.

Let's normalize our dataset before using it for analysis:

```
In [ ]:
        filled_df_components = df_components.fillna(method='ffill')
        daily_df_components = filled_df_components.resample('24h').ffill()
        daily_df_components = daily_df_components.fillna(method='bfill')
```

If you inspect every value in this data feed, you will notice NaN values, or missing data. Since we are using data that is error-prone, and for quick studies of PCA, we can temporarily fill in these unknown variables by propagating previous observed values. The `fillna(method='ffill')` method helps to do this and stores the result in the `filled_df_components` variable.

An additional step in normalizing is to resample the time series at regular intervals and match it up exactly with our Dow time series dataset, which we will be downloading later. The `daily_df_components` variable stores the result from resampling the time series on a daily basis, and any missing values during resampling are propagated using the forward fill method. And finally, to account for incomplete starting data, we will simply perform a backfill of values with `fillna(method='bfill')`.

> For the purpose of PCA demonstration, we have to make do with free, low-quality datasets. If you require high quality datasets, consider subscribing to a data publisher.

Quandl doesn't provide free datasets on the DJIA. In the next section, we will explore another data provider named Alpha Vantage as an alternative method of downloading datasets.

About Alpha Vantage

Alpha Vantage (`https://www.alphavantage.co`) is a data provider that provides real-time and historical data on equities, foreign exchange, and cryptocurrencies. Similar to Quandl, you can obtain a Python wrapper for the Alpha Vantage REST API interface and download free datasets directly into a `pandas` DataFrame.

Obtaining an Alpha Vantage API key

From your web browser, visit `https://www.alphavantage.co`, and click **Get your free API Key today** from the home page. You will be brought to a registration page. Fill in basic information about yourself and submit the form. Your API key will be shown in the same page. Copy this API key for use in the next section:

Installing the Alpha Vantage Python wrapper

From your terminal window, type the following command to install the Python module for Alpha Vantage:

```
$ pip install alpha_vantage
```

Downloading the DJIA dataset from Alpha Vantage

The following code connects to Alpha Vantage and downloads the Dow dataset, with the ticker code ^DJI. Replace the value of the constant variable, ALPHA_VANTAGE_API_KEY, with your own API key:

```
In [ ]:
    """
    Download the all-time DJIA dataset
    """
    from alpha_vantage.timeseries import TimeSeries

    # Update your Alpha Vantage API key here...
    ALPHA_VANTAGE_API_KEY = 'PZ2ISG9CYY379KLI'

    ts = TimeSeries(key=ALPHA_VANTAGE_API_KEY, output_format='pandas')
    df, meta_data = ts.get_daily_adjusted(symbol='^DJI', outputsize='full')
```

The TimeSeries class of the alpha_vantage.timeseries module is instantiated with the API key and specifies that datasets are automatically downloaded as pandas DataFrame objects. The get_daily_adjusted() method with the outputsize='full' parameter downloads the entire available daily adjusted prices for the given ticker symbol in the df variable as a DataFrame object.

Let's inspect this DataFrame with the info() command:

```
In [ ]:
    df.info()
Out[ ]:
    <class 'pandas.core.frame.DataFrame'>
    Index: 4760 entries, 2000-01-03 to 2018-11-30
    Data columns (total 8 columns):
    1. open                4760 non-null float64
    2. high                4760 non-null float64
    3. low                 4760 non-null float64
    4. close               4760 non-null float64
    5. adjusted close      4760 non-null float64
    6. volume              4760 non-null float64
    7. dividend amount     4760 non-null float64
    8. split coefficient   4760 non-null float64
    dtypes: float64(8)
    memory usage: 316.1+ KB
```

The Dow dataset that we downloaded from Alpha Vantage gives us the full time series data from the most recent available trading date, all the way back to the year 2000. It contains several columns that give us additional information.

Let's also inspect the indexes of this DataFrame:

```
In [ ]:
    df.index
Out[ ]:
    Index(['2000-01-03', '2000-01-04', '2000-01-05', '2000-01-06',
'2000-01-07',
            '2000-01-10', '2000-01-11', '2000-01-12', '2000-01-13',
'2000-01-14',
            ...
            '2018-08-17', '2018-08-20', '2018-08-21', '2018-08-22',
'2018-08-23',
            '2018-08-24', '2018-08-27', '2018-08-28', '2018-08-29',
'2018-08-30'],
           dtype='object', name='date', length=4696)
```

The outputs suggests that the index values are made up of an object of string type. Let's convert this DataFrame into something suitable for our analysis:

```
In [ ]:
    import pandas as pd

    # Prepare the dataframe
    df_dji = pd.DataFrame(df['5. adjusted close'])
    df_dji.columns = ['DJIA']
    df_dji.index = pd.to_datetime(df_dji.index)

    # Trim the new dataframe and resample
    djia_2017 = pd.DataFrame(df_dji.loc['2017-01-01':'2017-12-31'])
    djia_2017 = djia_2017.resample('24h').ffill()
```

Here, we are taking the adjusted closing prices of Dow Jones for the year of 2017, resampled on a daily basis. The resulting DataFrame object is stored in `djia_2017`, which we can use for applying PCA.

Applying a kernel PCA

In this section, we will perform kernel PCA to find eigenvectors and eigenvalues so that we can reconstruct the Dow index.

Finding eigenvectors and eigenvalues

We can perform a kernel PCA using the `KernelPCA` class of the `sklearn.decomposition` module in Python. The default kernel method is linear. The dataset that's used in PCA is required to be normalized, which we can perform with z-scoring. The following code do this:

```
In [ ]:
    from sklearn.decomposition import KernelPCA

    fn_z_score = lambda x: (x - x.mean()) / x.std()

    df_z_components = daily_df_components.apply(fn_z_score)
    fitted_pca = KernelPCA().fit(df_z_components)
```

The `fn_z_score` variable is an inline function to perform z-scoring on a `pandas` DataFrame, which is applied with the `apply()` method. These normalized datasets can be fitted into a kernel PCA with the `fit()` method. The fitted results of the daily Dow component prices are stored in the `fitted_pca` variable, which is of the same `KernelPCA` object.

Two main outputs of PCA are eigenvectors and eigenvalues. **Eigenvectors** are vectors containing the direction of the principal component line, which doesn't change when a linear transformation is applied. **Eigenvalues** are scalar values indicating the amount of variance of the data in a direction with respect to a particular eigenvector. In fact, the eigenvector with the highest eigenvalue forms the principal component.

The `alphas_` and `lambdas_` attributes of the `KernelPCA` object return the eigenvectors and eigenvalues of the centered kernel matrix dataset, respectively. When we plot the eigenvalues, we get the following:

```
In [ ]:
    %matplotlib inline
    import matplotlib.pyplot as plt

    plt.rcParams['figure.figsize'] = (12,8)
    plt.plot(fitted_pca.lambdas_)
    plt.ylabel('Eigenvalues')
    plt.show();
```

We should then get the following output:

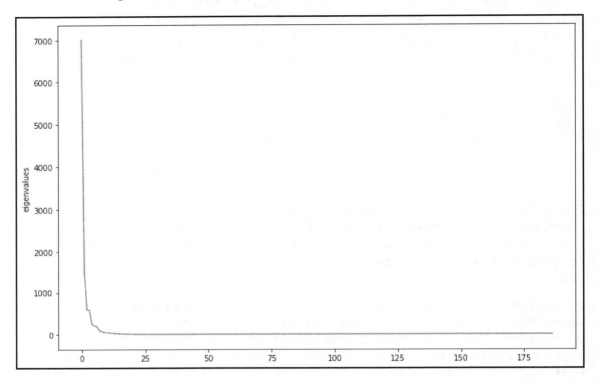

We can see that the first few eigenvalues explain much of the variances in the data, and become more negligent further down the components. Taking the first five eigenvalues, let's see how much explanation each of these eigenvalues gives us by obtaining their weighted average values:

```
In [ ]:
    fn_weighted_avg = lambda x: x / x.sum()
    weighted_values = fn_weighted_avg(fitted_pca.lambdas_)[:5]
In [ ]:
    print(weighted_values)
Out[ ]:
    array([0.64863002, 0.13966718, 0.05558246, 0.05461861, 0.02313883])
```

We can see that the first component explains 65% of the variance of the data, the second component explains 14%, and so on. Taking the sum of these values, we get the following:

```
In [ ]:
    weighted_values.sum()
Out[ ]:
    0.9216371041932268
```

The first five eigenvalues would explain 92% of the variance in the dataset.

Reconstructing the Dow index with PCA

By default, the `KernelPCA` is instantiated with the `n_components=None` parameter, which constructs a kernel PCA with non-zero components. We can also create a PCA index with five components:

```
In [ ]:
    import numpy as np

    kernel_pca = KernelPCA(n_components=5).fit(df_z_components)
    pca_5 = kernel_pca.transform(-daily_df_components)

    weights = fn_weighted_avg(kernel_pca.lambdas_)
    reconstructed_values = np.dot(pca_5, weights)

    # Combine DJIA and PCA index for comparison
    df_combined = djia_2017.copy()
    df_combined['pca_5'] = reconstructed_values
    df_combined = df_combined.apply(fn_z_score)
    df_combined.plot(figsize=(12, 8));
```

With the `fit()` method, we fitted the normalized dataset using the linear kernel PCA function with five components. The `transform()` method transforms the original dataset with the kernel PCA. These values are normalized using the weights indicated by the eigenvectors, computed with dot matrix multiplication. We then create a copy of the Dow time series `pandas` DataFrame with the `copy()` method, and combine it with the reconstructed values in the `df_combined` DataFrame.

The new DataFrame is normalized by z-scoring, and plotted out to see how well the reconstructed PCA index tracks the original Dow movements. This gives us the following output:

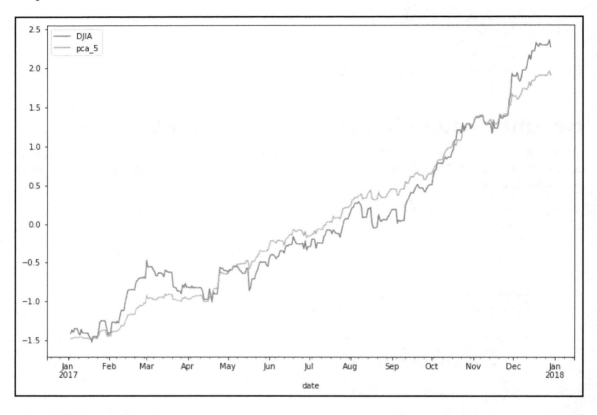

The preceding graph shows the original Dow index against the reconstructed Dow index with five principal components for the year 2017.

Stationary and non-stationary time series

It is important that time series data that's used for statistical analysis is stationary in order to perform statistical modeling correctly, as such usages may be for prediction and forecasting. This section introduces the concepts of stationarity and non-stationarity in time series data.

Stationarity and non-stationarity

In empirical time series studies, price movements are observed to drift toward some long-term mean, either upwards or downwards. A stationary time series is one whose statistical properties, such as mean, variance, and autocorrelation, are constant over time. Conversely, observations on non-stationary time series data have their statistical properties change over time, mostly likely due to trends, seasonality, presence of a unit root, or a combination of all three.

In time series analysis, it is assumed that the data of the underlying process is stationary. Otherwise, modeling from non-stationary data may produce unpredictable results. This would lead to a condition known as spurious regression. **Spurious regression** is a regression that produces misleading statistical evidence of relationships between independent non-stationary variables. In order to receive consistent and reliable results, non-stationary data needs to be transformed into stationary data.

Checking for stationarity

There are a number of ways to check whether time series data is stationary or non-stationary:

- **Through visualizations**: You can review a time series graph for obvious indication of trends or seasonality.
- **Through statistical summaries**: You can review the statistical summaries of your data significant differences. For example, you can partition your time series data and compare the mean and variance of each group.
- **Through statistical tests**: You can use statistical tests such as the Augmented Dickey-Fuller Test to check if stationarity expectations have been met or violated.

Types of non-stationary processes

The following points help to identify non-stationary behavior in time series data for consideration in transforming stationary data:

- **Pure random walk**: A process with a unit root or a stochastic trend. It is a non-mean reverting process with a variance that evolves over time and goes to infinity.
- **Random walk with drift**: A process with a random walk and a constant drift.

- **Deterministic trend**: A process with a mean that grows around a fixed trend, which is constant and independent of time.
- **Random walk with drift and deterministic trend**: A process combining a random walk with a drift component, and a deterministic trend.

Types of stationary processes

These are a number of definitions of stationarity that you may come across in time series studies:

- **Stationary process**: A process that generates a stationary series of observations.
- **Trend stationary**: A process that does not exhibit a trend.
- **Seasonal stationary**: A process that does not exhibit seasonality.
- **Strictly stationary**: Also known as **strongly stationary**. A process whose unconditional joint probability distribution of random variables does not change when shifted in time (or along the x axis).
- **Weakly stationary**: Also known as **covariance-stationary**, or **second-order stationary**. A process whose mean, variance, and correlation of random variables doesn't change when shifted in time.

The Augmented Dickey-Fuller Test

An **Augmented Dickey-Fuller Test** (**ADF**) is a type of statistical test that determines whether a unit root is present in time series data. Unit roots can cause unpredictable results in time series analysis. A null hypothesis is formed on the unit root test to determine how strongly time series data is affected by a trend. By accepting the null hypothesis, we accept the evidence that the time series data is non-stationary. By rejecting the null hypothesis, or accepting the alternative hypothesis, we accept the evidence that the time series data is generated by a stationary process. This process is also known as **trend-stationary.** Values of the ADF test statistic are negative. Lower values of ADF indicates stronger rejection of the null hypothesis.

Here are some basic autoregression models for use in ADF testing:

- No constant and no trend:

$$\Delta y_t = \gamma y_{t-1} + \sum_{j=1}^{p} \delta_j \Delta y_{t-j} + \epsilon_t$$

- A constant without a trend:

$$\Delta y_t = \alpha + \gamma y_{t-1} + \sum_{j=1}^{p} \delta_j \Delta y_{t-j} + \epsilon_t$$

- With a constant and trend:

$$\Delta y_t = \alpha + \gamma y_{t-1} + \beta t + \sum_{j=1}^{p} \delta_j \Delta y_{t-j} + \epsilon_t$$

Here, α is the drift constant, β is the coefficient on a time trend, γ is the coefficient of our hypothesis, p is the lag order of the first-differences autoregressive process, and ϵ_t is an independent and identically distributed residual term. When $\alpha=0$ and $\beta=0$, the model is a random walk process. When $\beta=0$, the model is a random walk with a drift process. The length of the lag p is to be chosen so that the residuals are not serially correlated. Some approaches for examining the information criteria for choosing lags are by minimizing the **Akaike information criterion** (**AIC**), the **Bayesian information criterion** (**BIC**), and the **Hannan-Quinn information criterion**.

The hypothesis can then be formulated as follows:

- Null hypothesis, H_0: If failed to be rejected, it suggests that the time series contains a unit root and is non-stationary
- Alternate hypothesis, H_1: If H_0 is rejected, it suggests that the time series does not contain a unit root and is stationary

To accept or reject the null hypothesis, we use the p-value. We reject the null hypothesis if the p-value falls below a threshold value such as 5% or even 1%. We can fail to reject the null hypothesis if the p-value is above this threshold value and consider the time series as non-stationary. In other words, if our threshold value is 5%, or 0.05, note the following:

- p-value > 0.05: We fail to reject the null hypothesis H_0 and conclude that the data has a unit root and is non-stationary
- p-value ≤ 0.05: We reject the null hypothesis H_0 and conclude that the data has a unit root and is non-stationary

The `statsmodels` library provides the `adfuller()` function that implements this test.

Analyzing a time series with trends

Let's examine a time series dataset. Take, for example, the prices of gold futures traded on the CME. On Quandl, the gold futures continuous contract is available for download with the following code: `CHRIS/CME_GC1`. This data is curated by the Wiki Continuous Futures community group, taking into account the front month contracts only. The sixth column of the dataset contains the settlement prices. The following code downloads the dataset from the year 2000 onward:

```
In [ ]:
    import quandl

    QUANDL_API_KEY = 'BCzkk3NDWt7H9yjzx-DY'  # Your Quandl key here
    quandl.ApiConfig.api_key = QUANDL_API_KEY

    df = quandl.get(
        'CHRIS/CME_GC1',
        column_index=6,
        collapse='monthly',
        start_date='2000-01-01')
```

Examine the head of the dataset using the following command:

```
In [ ]:
    df.head()
```

We get the following table:

Settle	Date
2000-01-31	283.2
2000-02-29	294.2
2000-03-31	278.4
2000-04-30	274.7
2000-05-31	271.7

Compute the rolling mean and standard deviation into the df_mean and df_std variables, respectively, with a window period of one year:

```
In [ ] :
    df_settle = df['Settle'].resample('MS').ffill().dropna()

    df_rolling = df_settle.rolling(12)
    df_mean = df_rolling.mean()
    df_std = df_rolling.std()
```

The resample() method helps to ensure that the data is smoothed out on a monthly basis, and the ffill() method forward fills any missing values.

> A list of useful common time series frequencies for specifying the resample() method can be found at http://pandas.pydata.org/pandas-docs/stable/timeseries.html#offset-aliases.

Let's visualize the plot of the rolling mean against the original time series:

```
In [ ] :
    plt.figure(figsize=(12, 8))
    plt.plot(df_settle, label='Original')
    plt.plot(df_mean, label='Mean')
    plt.legend();
```

We obtain the following output:

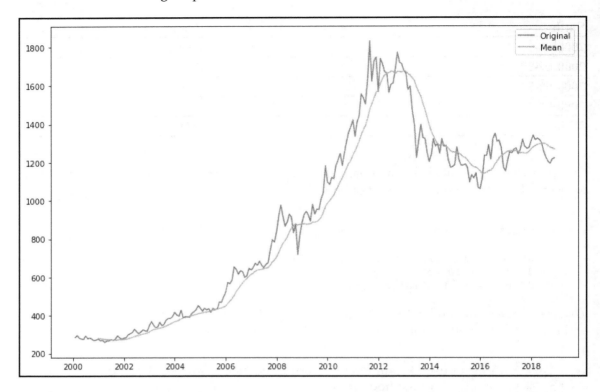

Visualizing the rolling standard deviation separately, we get the following:

```
In [ ] :
    df_std.plot(figsize=(12, 8));
```

We obtain the following output:

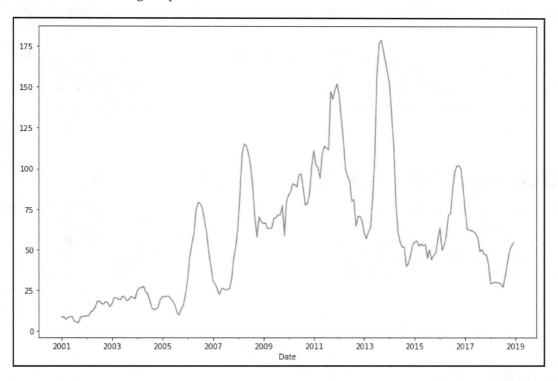

Using the `statsmodels` module, perform an ADF unit root test on our dataset with the `adfuller()` method:

```
In [ ]:
    from statsmodels.tsa.stattools import adfuller

    result = adfuller(df_settle)
    print('ADF statistic: ',  result[0])
    print('p-value:', result[1])

    critical_values = result[4]
    for key, value in critical_values.items():
        print('Critical value (%s): %.3f' % (key, value))
Out[ ]:
    ADF statistic:  -1.4017828015895548
    p-value: 0.5814211232134314
    Critical value (1%): -3.461
    Critical value (5%): -2.875
    Critical value (10%): -2.574
```

The `adfuller()` method returns a tuple of seven values. Particularly, we are interested in the first, second, and fifth values, which give us the test statistic, `p-value`, and a dictionary of critical values, respectively.

Observe from the plots that the mean and standard deviations swing over time, with the mean exhibiting an overall upward trend. The ADF test statistic value is more than the critical values (especially at 5%), and the `p-value` is more than 0.05. With these, we cannot reject the null hypothesis that there is a unit root and consider that our data is non-stationary.

Making a time series stationary

A non-stationary time series data is likely to be affected by a trend or seasonality. Trending time series data has a mean that is not constant over time. Data that is affected by seasonality have variations at specific intervals in time. In making a time series data stationary, the trend and seasonality effects have to be removed. Detrending, differencing, and decomposition are such methods. The resulting stationary data is then suitable for statistical forecasting.

Let's look at all three methods in detail.

Detrending

The process of removing a trend line from a non-stationary data is known as **detrending**. This involves a transformation step that normalizes large values into smaller ones. Examples could be a logarithmic function, a square root function, or even a cube root. A further step is to subtract the transformation from the moving average.

Let's perform detrending on the same dataset, `df_settle`, with logarithmic transformation and subtracting from the moving average of two periods, as given in the following Python code:

```
In [ ]:
    import numpy as np

    df_log = np.log(df_settle)
In [ ]:
    df_log_ma= df_log.rolling(2).mean()
    df_detrend = df_log - df_log_ma
    df_detrend.dropna(inplace=True)
```

```
# Mean and standard deviation of detrended data
df_detrend_rolling = df_detrend.rolling(12)
df_detrend_ma = df_detrend_rolling.mean()
df_detrend_std = df_detrend_rolling.std()

# Plot
plt.figure(figsize=(12, 8))
plt.plot(df_detrend, label='Detrended')
plt.plot(df_detrend_ma, label='Mean')
plt.plot(df_detrend_std, label='Std')
plt.legend(loc='upper right');
```

The df_log variable is our transformed pandas DataFrame by logarithmic function using the numpy module, and the df_detrend variable contains the detrended data. We plot this detrended data to visualize its mean and standard deviation over a rolling one-year period.

We get the following output:

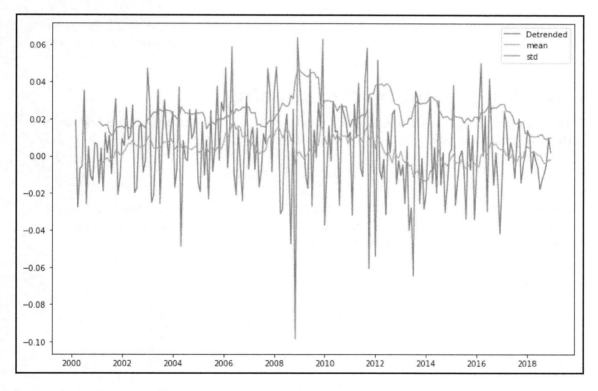

Observe that the mean and standard deviation do not exhibit a long-term trend.

Looking at the ADF test statistic for the detrended data, we get the following:

```
In [ ]:
    from statsmodels.tsa.stattools import adfuller

    result = adfuller(df_detrend)
    print('ADF statistic: ', result[0])
    print('p-value: %.5f' % result[1])

    critical_values = result[4]
    for key, value in critical_values.items():
        print('Critical value (%s): %.3f' % (key, value))
Out [ ]:
    ADF statistic:  -17.04239232215001
    p-value: 0.00000
    Critical value (1%): -3.460
    Critical value (5%): -2.874
    Critical value (10%): -2.574
```

The `p-value` for this detrended data is less than 0.05. Our ADF test statistic is lower than all the critical values. We can reject the null hypothesis and say that this data is stationary.

Removing trend by differencing

Differencing involves the difference of time series values with a time lag. The first-order difference of the time series is given by the following formula:

$$\Delta y_t = \log(y_t) - \log(y_{t-1})$$

We can reuse the `df_log` variable in the previous section as our logarithmic transformed time series, and utilize the `diff()` and `shift()` methods of NumPy modules in our differencing, with the following code:

```
In [ ]:
    df_log_diff = df_log.diff(periods=3).dropna()

    # Mean and standard deviation of differenced data
    df_diff_rolling = df_log_diff.rolling(12)
    df_diff_ma = df_diff_rolling.mean()
    df_diff_std = df_diff_rolling.std()

    # Plot the stationary data
    plt.figure(figsize=(12, 8))
    plt.plot(df_log_diff, label='Differenced')
    plt.plot(df_diff_ma, label='Mean')
```

```
plt.plot(df_diff_std, label='Std')
plt.legend(loc='upper right');
```

The parameter of `diff()` given as `periods=3` indicates that the dataset is shifted by three periods in calculating the differences.

This provides the following output:

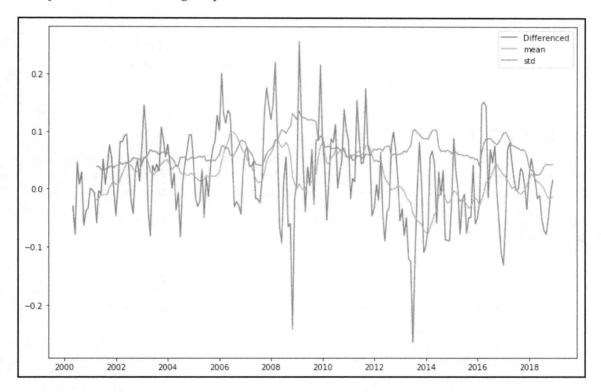

Observe from the plots that the rolling mean and standard deviation tend to change very little over time.

Looking at our ADF test statistic, we get the following:

```
In [ ]:
    from statsmodels.tsa.stattools import adfuller

    result = adfuller(df_log_diff)

    print('ADF statistic:', result[0])
    print('p-value: %.5f' % result[1])
```

```
    critical_values = result[4]
    for key, value in critical_values.items():
        print('Critical value (%s): %.3f' % (key, value))
Out[ ]:
    ADF statistic: -2.931684356800213
    p-value: 0.04179
    Critical value (1%): -3.462
    Critical value (5%): -2.875
    Critical value (10%): -2.574
```

From the ADF test, the `p-value` for this data is less than 0.05. Our ADF test statistic is lower than the 5% critical value, indicating a 95% confidence level that this data is stationary. We can reject the null hypothesis and say that this data is stationary.

Seasonal decomposing

Decomposing involves modeling both the trend and seasonality, and then removing them. We can use the `statsmodel.tsa.seasonal` module to model a non-stationary time series dataset using moving averages and remove its trend and seasonal components.

By reusing our `df_log` variable containing the logarithm of our dataset from the previous section, we get the following:

```
In [ ]:
    from statsmodels.tsa.seasonal import seasonal_decompose

    decompose_result = seasonal_decompose(df_log.dropna(), freq=12)

    df_trend = decompose_result.trend
    df_season = decompose_result.seasonal
    df_residual = decompose_result.resid
```

The `seasonal_decompose()` method of `statsmodels.tsa.seasonal` requires a parameter, `freq`, which is an integer value specifying the number of periods per seasonal cycle. Since we are using monthly data, we expect 12 periods in a seasonal year. The method returns an object with three attributes, mainly the trend and seasonal components, as well as the final `pandas` series data with its trend and seasonal components removed.

More information on the `seasonal_decompose()` method of the `statsmodels.tsa.seasonal` module can be found at `https://www.statsmodels.org/dev/generated/statsmodels.tsa.seasonal.seasonal_decompose.html`.

Let's visualize the different plots by running the following Python code:

```
In [ ]:
    plt.rcParams['figure.figsize'] = (12, 8)
    fig = decompose_result.plot()
```

We get the following graphs:

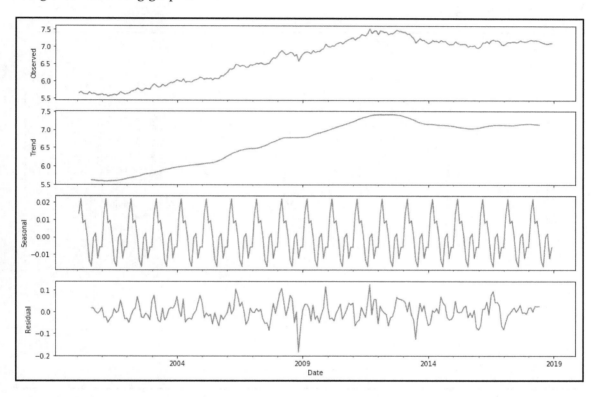

Here, we can see the individual trend and seasonality components being removed from the dataset and plotted, and the residuals plotted at the bottom. Let's visualize the statistical properties of our residuals:

```
In [ ]:
    df_log_diff = df_residual.diff().dropna()

    # Mean and standard deviation of differenced data
    df_diff_rolling = df_log_diff.rolling(12)
    df_diff_ma = df_diff_rolling.mean()
    df_diff_std = df_diff_rolling.std()
```

```
# Plot the stationary data
plt.figure(figsize=(12, 8))
plt.plot(df_log_diff, label='Differenced')
plt.plot(df_diff_ma, label='Mean')
plt.plot(df_diff_std, label='Std')
plt.legend();
```

We get the following graph:

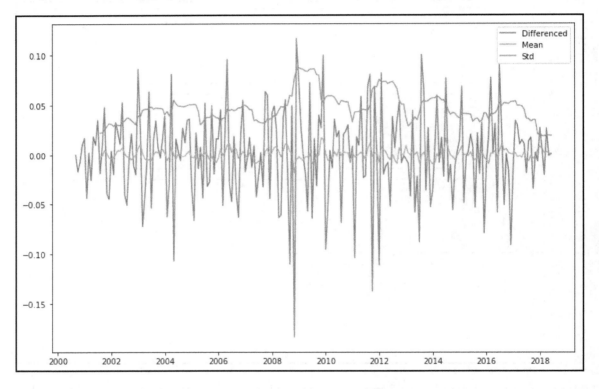

Observe from the plots that the rolling mean and standard deviation tend to change very little over time.

By checking our residual data for stationarity, we get the following:

```
In [ ]:
    from statsmodels.tsa.stattools import adfuller

    result = adfuller(df_residual.dropna())

    print('ADF statistic:',  result[0])
    print('p-value: %.5f' % result[1])
```

```
        critical_values = result[4]
        for key, value in critical_values.items():
            print('Critical value (%s): %.3f' % (key, value))
Out[ ]:
        ADF statistic: -6.468683205304995
        p-value: 0.00000
        Critical value (1%): -3.463
        Critical value (5%): -2.876
        Critical value (10%): -2.574
```

From the ADF test, the `p-value` for this data is less than 0.05. Our ADF test statistic is lower than all the critical values. We can reject the null hypothesis and say that this data is stationary.

Drawbacks of ADF testing

Here are some considerations when using ADF tests for reliable checking of non-stationary data:

- The ADF test do not truly tell apart between pure and non-unit root generating processes. In long-term moving average processes, the ADF tests becomes biased in rejecting the null hypothesis. Other stationarity testing methods such as the **Kwiatkowski–Phillips–Schmidt–Shin (KPSS)** tests and the **Phillips-Perron** test take a different approach in treating the presence of unit roots.
- There is no fixed methodology in determining the lag length p. If p is too small, the remaining serial correlation in the errors may affect the size of the test. If p is too large, the power of the test will deteriorate. Additional consideration is to be given for this lag order.
- As deterministic terms are added to the test regressions, the power of unit root tests diminishes.

Forecasting and predicting a time series

In the previous section, we identified non-stationarity in time series data and discussed techniques for making time series data stationary. With stationary data, we can proceed to perform statistical modeling such as prediction and forecasting. Prediction involves generating best estimates of in-sample data. Forecasting involves generating best estimates of out-of-sample data. Predicting future values is based on previously observed values. One such commonly used method is the Autoregressive Integrated Moving Average.

About the Autoregressive Integrated Moving Average

The **Autoregressive Integrated Moving Average** (**ARIMA**) is a forecasting model for stationary time series based on linear regression. As its name suggests, it is based on three components:

- **Autoregression** (**AR**): A model that uses the dependency between an observation and its lagged values
- **Integrated** (**I**): The use of differencing an observation with an observation from a previous time stamp in making the time series stationary
- **Moving average** (**MA**): A model that uses the dependency between an observed error term and a combination of previous error terms, e_t

ARIMA models are referenced by the notation *ARIMA(p, d, q)*, which corresponds to the parameters of the three components. Non-seasonal ARIMA models can be specified by changing the values of *p*, *d*, and *q*, as follows:

- **ARIMA(*p*,0,0)**: First-order autoregressive model, notated by *AR(p)*. *p* is the lag order, indicating the number of lagged observations in the model. For example, *ARIMA(2,0,0)* is *AR(2)* and represented as follows:

$$Y_t = c + \phi y_{t-1} + \phi y_{t-2} + e_t$$

    Here, ϕ_1 and ϕ_2 are parameters for the model.

- **ARIMA(0,*d*,0)**: First degree of differencing in the integrated component, also known as random walk, notated by *I(d)*. *d* is the degree of differencing, indicating the number of times the data have had past values subtracted. For example, *ARIMA(0,1,0)* is *I(1)* and represented as follows:

$$Y_t = Y_{t-1} + u$$

    Here, μ is the mean of the seasonal difference.

- **ARIMA(0,0,*q*)**: Moving average component, notated by *MA(q)*. The order *q* determines the number of terms to be included in the model:

$$Y_t = c + \phi_1 e_{t-1} + \phi_2 e_{t-2} + \ldots + \phi_q e_{t-q} + e_t$$

Finding model parameters by grid search

A grid search, also known as the hyperparameter optimization method, can be used to iteratively explore different combinations of parameters for fitting our ARIMA model. We can fit a seasonal ARIMA model with the `SARIMAX()` function of the `statsmodels` module in each iteration, returning an object of the `MLEResults` class. The `MLEResults` object holds an `aic` attribute for returning the AIC value. The model with the lowest AIC value gives us the best-fitting model that determines our parameters of *p*, *d*, and *q*. More information on SARIMAX can be found at `https://www.statsmodels.org/dev/generated/statsmodels.tsa.statespace.sarimax.SARIMAX.html`.

We define the grid search procedure as the `arima_grid_search()` function, as follows:

```
In [ ]:
    import itertools
    import warnings
    from statsmodels.tsa.statespace.sarimax import SARIMAX

    warnings.filterwarnings("ignore")

    def arima_grid_search(dataframe, s):
        p = d = q = range(2)
        param_combinations = list(itertools.product(p, d, q))
        lowest_aic, pdq, pdqs = None, None, None
        total_iterations = 0
        for order in param_combinations:
            for (p, q, d) in param_combinations:
                seasonal_order = (p, q, d, s)
                total_iterations += 1
                try:
                    model = SARIMAX(df_settle, order=order,
                        seasonal_order=seasonal_order,
                        enforce_stationarity=False,
                        enforce_invertibility=False,
                        disp=False
                    )
                    model_result = model.fit(maxiter=200, disp=False)

                    if not lowest_aic or model_result.aic < lowest_aic:
                        lowest_aic = model_result.aic
                        pdq, pdqs = order, seasonal_order

                except Exception as ex:
                    continue

        return lowest_aic, pdq, pdqs
```

Our variable, `df_settle`, holds the monthly prices of the futures data that we downloaded in the previous section. In the **SARIMAX (seasonal autoregressive integrated moving average with exogenous regressors model)** function, we provided the `seasonal_order` parameter, which is the *ARIMA(p,d,q,s)* seasonal component, where *s* is the number of periods in a season of the dataset. Since we are using monthly data, we use 12 periods to define a seasonal pattern. The `enforce_stationarity=False` parameter doesn't transform the AR parameters to enforce stationarity in the AR component of the model. The `enforce_invertibility=False` parameter doesn't transform MA parameters to enforce invertibility in the MA component of the model. The `disp=False` parameter suppresses output information when fitting our models.

With the grid function defined, we can now call this with our monthly data and print out the model parameters with the lowest AIC value:

```
In [ ]:
    lowest_aic, order, seasonal_order = arima_grid_search(df_settle, 12)
In [ ]:
    print('ARIMA{}x{}'.format(order, seasonal_order))
    print('Lowest AIC: %.3f'%lowest_aic)
Out[ ]:
    ARIMA(0, 1, 1)x(0, 1, 1, 12)
    Lowest AIC: 2149.636
```

An `ARIMA(0,1,1,12)` seasonal component model would give us the lowest AIC value at 2149.636. We shall use these parameters to fit our SARIMAX model in the next section.

Fitting the SARIMAX model

Having obtained the optimal model parameters, inspect the model properties using the `summary()` method on the fitted results to view detailed statistical information:

```
In [ ]:
    model = SARIMAX(
        df_settle,
        order=order,
        seasonal_order=seasonal_order,
        enforce_stationarity=False,
        enforce_invertibility=False,
        disp=False
    )

    model_results = model.fit(maxiter=200, disp=False)
    print(model_results.summary())
```

This gives us the following output:

```
                       Statespace Model Results
================================================================================
===============
Dep. Variable:                        Settle   No. Observations:
226
Model:            SARIMAX(0, 1, 1)x(0, 1, 1, 12)   Log Likelihood
-1087.247
Date:                         Sun, 02 Dec 2018   AIC
2180.495
Time:                                 17:38:32   BIC
2190.375
Sample:                             02-01-2000   HQIC
2184.494
                                  - 11-01-2018
Covariance Type:                           opg
================================================================================
===
                 coef    std err          z      P>|z|      [0.025
0.975]
--------------------------------------------------------------------------------
---
ma.L1         -0.1716      0.044     -3.872      0.000      -0.258
-0.085
ma.S.L12      -1.0000    447.710     -0.002      0.998    -878.496
876.496
sigma2      2854.6342   1.28e+06      0.002      0.998      -2.5e+06
2.51e+06
================================================================================
========
Ljung-Box (Q):                          67.93   Jarque-Bera (JB):
52.74
Prob(Q):                                 0.00   Prob(JB):
0.00
Heteroskedasticity (H):                  6.98   Skew:
-0.34
Prob(H) (two-sided):                     0.00   Kurtosis:
5.43
================================================================================
========

Warnings:
[1] Covariance matrix calculated using the outer product of gradients
(complex-step).
```

It is important to run model diagnostics to investigate that model assumptions haven't been violated:

```
In [ ]:
    model_results.plot_diagnostics(figsize=(12, 8));
```

We get the following output:

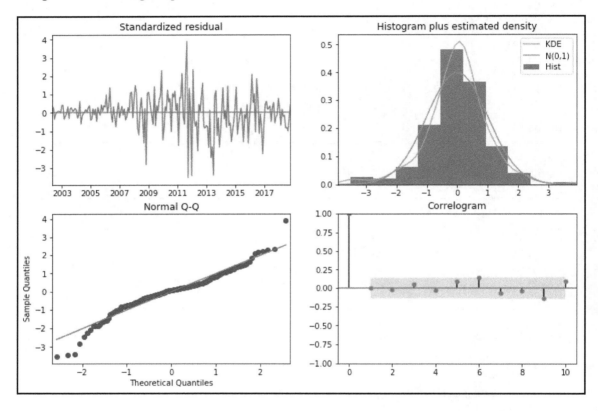

The top-right plot shows the **kernel density estimate** (**KDE**) of the standardized residuals, which suggests the errors are Gaussian with a mean close to zero. Let's see a more accurate statistic of the residuals:

```
In [ ] :
    model_results.resid.describe()
Out[ ]:
    count    223.000000
    mean       0.353088
    std       57.734027
    min     -196.799109
```

```
25%         -22.036234
50%           3.500942
75%          22.872743
max         283.200000
dtype: float64
```

From the description of the residuals, the non-zero mean suggests that the prediction may be biased positively.

Predicting and forecasting the SARIMAX model

The `model_results` variable is a `SARIMAXResults` object of the `statsmodel` module, representing the output of the SARIMAX model. It contains a `get_prediction()` method for performing in-sample prediction and out-of-sample forecasting. It also contains a `conf_int()` method, which returns the confidence intervals of the predictions, both lower- and upper-bounded, of the fitted parameters, which is at a 95% confidence interval by default. Let's apply these methods:

```
In [ ]:
    n = len(df_settle.index)
    prediction = model_results.get_prediction(
        start=n-12*5,
        end=n+5
    )
    prediction_ci = prediction.conf_int()
```

The `start` parameter in the `get_prediction()` method indicates we are performing an in-sample prediction of the most recent five years' prices. At the same time, with the `end` parameter, we are performing an out-of-sample forecast of the next five months.

By inspecting the top three forecasted confidence interval values, we get the following:

```
In [ ]:
    print(prediction_ci.head(3))
Out[ ]:
                lower Settle   upper Settle
    2017-09-01   1180.143917    1396.583325
    2017-10-01   1204.307842    1420.747250
    2017-11-01   1176.828881    1393.268289
```

Let's plot the predicted and forecasted prices against our original dataset, from the year 2008 onwards:

```
In  [ ]:
    plt.figure(figsize=(12, 6))

    ax = df_settle['2008':].plot(label='actual')
    prediction_ci.plot(
        ax=ax, style=['--', '--'],
        label='predicted/forecasted')

    ci_index = prediction_ci.index
    lower_ci = prediction_ci.iloc[:, 0]
    upper_ci = prediction_ci.iloc[:, 1]

    ax.fill_between(ci_index, lower_ci, upper_ci,
        color='r', alpha=.1)

    ax.set_xlabel('Time (years)')
    ax.set_ylabel('Prices')

    plt.legend()
    plt.show()
```

This gives us the following output:

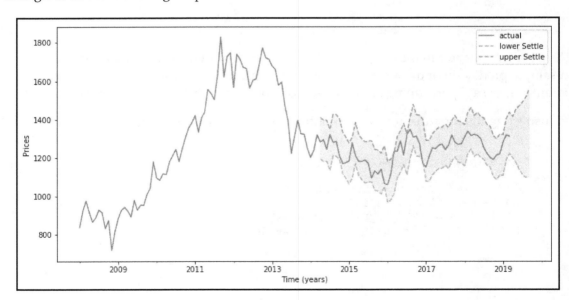

The solid line plot shows the observed values, while the dotted lines plot the five-year rolling predictions trailing closely and bounded by the confidence intervals in the shaded area. Observe that as the next five-month forecast goes into the future, and the confidence interval widens to reflect the loss of certainty in the outlook.

Summary

In this chapter, we were introduced to PCA as a dimension reduction technique in portfolio modeling. By breaking down the movement of asset prices of a portfolio into its principal components, or common factors, the most useful factors can be kept, and portfolio analysis can be greatly simplified without compromising on computational time and space complexity. In applying PCA to the Dow and its thirty components using the `KernelPCA` function of the `sklearn.decomposition` module, we obtained eigenvectors and eigenvalues, which we used to reconstruct the Dow with five components.

In the statistical analysis of time series data, the data is considered as either stationary or non-stationary. Stationary time series data is data whose statistical properties are constant over time. Non-stationary time series data has its statistical properties change over time, most likely due to trends, seasonality, presence of a unit root, or a combination of all three. Modeling from non-stationary data may produce spurious regression. In order to receive consistent and reliable results, non-stationary data needs to be transformed into stationary data.

We used statistical tests such as the ADF to check whether stationary expectations are met or violated. The `adfuller` method of the `statsmodels.tsa.stattools` module provides the test statistic, p-value, and critical values, from which we can fail to reject the null hypothesis that the data has a unit root and is non-stationary.

We transformed non-stationary data into stationary data by detrending, differencing, and seasonal decomposition. By using ARIMA, we fitted models using the `SARIMAX` function of the `statsmodels.tsa.statespace.sarimax` module to find suitable model parameters that give the lowest AIC value through an iterative grid search procedure. The fitted results are used for prediction and forecasting.

In the next chapter, we will perform interactive financial analytics with the VIX.

Section 3: A Hands-On Approach 3

In this section, we will apply the theoretical concepts covered in `Section 1`, *Getting Started with Python* and `Section 2`, *Financial Concepts* to build fully-functional working systems.

This section will contain the following chapters:

- `Chapter 7`, *Interactive Financial Analytics with VIX*
- `Chapter 8`, *Building an Algorithmic Trading Platform*
- `Chapter 9`, *Implementing a Backtesting System*
- `Chapter 10`, *Machine Learning for Finance*
- `Chapter 11`, *Deep Learning for Finance*

Interactive Financial Analytics with the VIX

7

Investors use volatility derivatives to diversify and hedge their risk in equity and credit portfolios. Since long-term investors in equity funds are exposed to downside risk, volatility can be used as a hedge for the tail risk and replacement for the put options. In the United States, the **Chicago Board Options Exchange (CBOE) Volatility Index (VIX)**, or simply called the **VIX**, measures the short-term volatility implied by S&P 500 stock index options with an average expiration of 30 days. Many people around the world use the VIX to measure stock market volatility over the next 30-day period. In Europe, the equivalent volatility counterpart indicator is the **EURO STOXX 50 Volatility (VSTOXX) Market Index**. For benchmark strategies utilizing the S&P 500 Index, the nature of its negative correlation with the VIX presents a viable way of avoiding benchmark-rebalancing costs. The statistical nature of volatility allows traders to perform mean-reverting strategies, dispersion trading, and volatility spread trading, among others.

In this chapter, we will take a look at how to perform data analytics on the VIX and the S&P 500 Index. Using the S&P 500 Index options, we can reconstruct the VIX and compare them with the observed values. The code presented here runs on the Jupyter Notebook, the interactive component of Python, to help us visualize data and study relationships between them.

In this chapter, we will discuss the following topics:

- An introduction to the EURO STOXX 50 index, VSTOXX, and the VIX
- Performing financial analytics on the S&P 500 Index and the VIX
- Step-by-step reconstruction of the VIX index in accordance with the CBOE VIX white paper
- Finding near-term and next-term options of the VIX index
- Determining strike price boundaries of options datasets
- Tabulating contributions to the VIX by strike prices

- Calculating forward levels of near-term and next-term options
- Calculating volatility values of near-term and next-term options
- Calculating multiple VIX indexes at once
- Comparing the results of the calculated index with the actual S&P 500 Index

Volatility derivatives

The two most popular volatility indexes worldwide are the VIX and VSTOXX, which are available in the United States and Europe respectively. The VIX is based on the S&P 500 Index that is disseminated on the CBOE. While the VIX is not traded directly, derivative products of the VIX such as options, futures, exchange-traded funds, and a host of volatility-based securities are available to investors. The CBOE website provides comprehensive information on many options and market indices such as the S&P 500 standard and weekly options, and the VIX, which we can analyze. We begin by understanding the background of these products before performing financial analytics on them in the later sections of this chapter.

STOXX and the Eurex

In the United States, the S&P 500 Index is one of the most widely watched stock market indexes, created by Standard & Poor's Dow Jones Indices. In Europe, one such company is STOXX Limited.

Formed in 1997, STOXX Limited is headquartered in Zurich, Switzerland and calculates approximately 7,000 indices globally. As an index provider, it develops, maintains, distributes, and markets a comprehensive range of indices that are known to be strictly rule-based and transparent.

STOXX provides a number of equity indices in these categories: benchmark indices, blue-chip indices, dividend indices, size indices, sector indices, style indices, optimized indices, strategy indices, theme indices, sustainability indices, faith-based indices, smart beta indices, and calculation products.

The Eurex Exchange is a derivatives exchange in Frankfurt, Germany offering more than 1,900 products, including equity indices, futures, options, ETFs, dividends, bonds, and repos. Many of STOXX's products and derivatives trade on the Eurex.

The EURO STOXX 50 Index

Designed by STOXX Limited, the EURO STOXX 50 Index is one of the most liquid stock indexes worldwide, serving many indices products listed on the Eurex. It was introduced on February 26, 1998 and is made up of 50 blue-chip stocks from the 12 Eurozone countries: Austria, Belgium, Finland, France, Germany, Greece, Ireland, Italy, Luxembourg, the Netherlands, Portugal, and Spain. The EURO STOXX 50 Index futures and options contracts are available and traded on the Eurex Exchange. Recalculation of the index takes place typically every 15 seconds based on real-time prices.

The ticker symbol for the EURO STOXX 50 Index is SX5E. EURO STOXX 50 Index Options take on the ticker symbol OESX.

The VSTOXX

The VSTOXX or EURO STOXX 50 Volatility is a class of volatility derivatives serviced by the Eurex Exchange. The VSTOXX Market Index is based on the market prices of a basket of OESX quoted **at the money** or **out of the money**. It measures the implied market volatility over the next 30 days on the EURO STOXX 50 Index.

Investors use volatility derivatives for benchmark strategies utilizing the EURO STOXX 50 Index; the nature of its negative correlation with the VSTOXX presents a viable way of avoiding benchmark-rebalancing costs. The statistical nature of volatility allows traders to perform mean-reverting strategies, dispersion trading, and volatility spread trading, among others. Recalculation of the index takes place every 5 seconds.

The ticker symbol for the VSTOXX is V2TX. VSTOXX Options and VSTOXX Mini Futures based on the VSTOXX Index trade on the Eurex Exchange.

The S&P 500 Index

The history of the **S&P 500 Index** (**SPX**) dates all the way back to 1923, where it was then known as the **Composite Index**. Initially, it tracked a small number of stocks. In 1957, the number of tracked stocks expanded to 500, and became the SPX.

Stocks that make up the SPX are publicly listed on the **New York Stock Exchange** (**NYSE**) or the **National Association of Securities Dealers Automated Quotations** (**NASDAQ**). The index is considered as a leading representation of the United States economy through large cap common stocks. Recalculation of the index takes place every 15 seconds and is distributed by Reuters America Holdings, Inc.

The common ticker symbols used by exchanges are *SPX* and *INX*, and on some websites is *^GSPC*.

SPX options

The CBOE offers a variety of options contracts to be traded, including options on stock indices such as the SPX. SPX Index options products come with different expiration dates. Standard or traditional SPX options expire on the third Friday of every month, and are settled at the start of the business. **SPX Weekly (SPXW)** options products may expire weekly, on Mondays, Wednesdays and Fridays, or monthly on the last trading day of the month. If the expiration date falls on an exchange holiday, the expiration date will be brought forward to the preceding business day. Other SPX options are the minis, trading at one-tenth of the notional size, and the **SPDR ETF (S&P's Depositary Receipt Exchange-traded Fund)**. Most SPX Index options are European style, with the exception of SPDR ETF which is American style.

The VIX

Like the STOXX, the CBOE VIX measures the short-term volatility implied by S&P 500 stock index option prices. The CBOE VIX began in 1993 based on the S&P 100 Index, was updated in 2003 to be based on the SPX, and updated again in 2014 to include SPXW options. Many people around the world think of the VIX to be a popular measurement tool for stock market volatility over the next 30-day period. The VIX recalculates every 15 seconds and is distributed by CBOE.

VIX Options and VIX Futures are based on the VIX and trade on the CBOE.

Financial analytics of the S&P 500 and the VIX

In this section, we will study the relationship between the VIX and the S&P 500 Market Index.

Gathering the data

We will be using Alpha Vantage as our data provider. Let's download the SPX and VIX datasets in the following steps:

1. Query for the all-time S&P 500 historical data with the ticker symbol ^GSPC:

```
In [ ]:
    from alpha_vantage.timeseries import TimeSeries

    # Update your Alpha Vantage API key here...
    ALPHA_VANTAGE_API_KEY = 'PZ2ISG9CYY379KLI'

    ts = TimeSeries(key=ALPHA_VANTAGE_API_KEY,
output_format='pandas')
    df_spx_data, meta_data = ts.get_daily_adjusted(
        symbol='^GSPC', outputsize='full')
```

2. Do the same for the VIX Index with the ticker symbol ^VIX:

```
In [ ]:
    df_vix_data, meta_data = ts.get_daily_adjusted(
        symbol='^VIX', outputsize='full')
```

3. Inspect the contents of the DataFrame object, df_spx_data:

```
In [ ]:
    df_spx_data.info()
Out[ ]:
    <class 'pandas.core.frame.DataFrame'>
    Index: 4774 entries, 2000-01-03 to 2018-12-21
    Data columns (total 8 columns):
    1. open                4774 non-null float64
    2. high                4774 non-null float64
    3. low                 4774 non-null float64
    4. close               4774 non-null float64
    5. adjusted close      4774 non-null float64
    6. volume              4774 non-null float64
    7. dividend amount     4774 non-null float64
    8. split coefficient   4774 non-null float64
    dtypes: float64(8)
    memory usage: 317.0+ KB
```

4. Inspect the contents of the DataFrame object, df_vix_data:

```
In [ ]:
    df_vix_data.info()
Out[ ]:
    <class 'pandas.core.frame.DataFrame'>
    Index: 4774 entries, 2000-01-03 to 2018-12-21
    Data columns (total 8 columns):
    1. open                4774 non-null float64
    2. high                4774 non-null float64
    3. low                 4774 non-null float64
    4. close               4774 non-null float64
    5. adjusted close      4774 non-null float64
    6. volume              4774 non-null float64
    7. dividend amount     4774 non-null float64
    8. split coefficient   4774 non-null float64
    dtypes: float64(8)
    memory usage: 317.0+ KB
```

5. Observe that the start date of both datasets begins from January 3, 2000, and that the fifth column labeled 5. adjusted close contains our values of interest. Extract these two columns and combine them into a single pandas DataFrame:

```
In [ ]:
    import pandas as pd

    df = pd.DataFrame({
        'SPX': df_spx_data['5. adjusted close'],
        'VIX': df_vix_data['5. adjusted close']
    })
    df.index = pd.to_datetime(df.index)
```

6. The last line to_datetime() method of pandas converts the trading dates given as string objects into a pandas DatetimeIndex object. Inspecting the head of our final DataFrame object df gives us the following:

```
In [ ]:
    df.head(3)
```

This gives us the following table:

Date	SPX	VIX
2000-01-03	1455.22	24.21
2000-01-04	1399.42	27.01
2000-01-05	1402.11	26.41

Viewing our formatted indexes shows us the following:

```
In [ ]:
    df.index
Out[ ]:
    DatetimeIndex(['2000-01-03', '2000-01-04', '2000-01-05', '2000-01-06',
                   '2000-01-07', '2000-01-10', '2000-01-11', '2000-01-12',
                   '2000-01-13', '2000-01-14',
                   ...
                   '2018-10-11', '2018-10-12', '2018-10-15', '2018-10-16',
                   '2018-10-17', '2018-10-18', '2018-10-19', '2018-10-22',
                   '2018-10-23', '2018-10-24'],
                  dtype='datetime64[ns]', name='date', length=4734,
freq=None)
```

With the `pandas` DataFrame formatted properly, let's proceed to work with this dataset.

Performing analytics

The `describe()` method of `pandas` gives us a summary statistic and distribution of values inside each column of the `pandas` DataFrame object:

```
In [ ]:
    df.describe()
```

This gives us the following table:

	SPX
count	4734.000000
mean	1493.538998
std	500.541938
min	676.530000
25%	1140.650000
50%	1332.730000

	SPX
75%	1840.515000
max	2930.750000

Another related method, info(), used earlier, gives us a technical summary of the DataFrame, such as the range of the index and memory usage:

```
In [ ]:
    df.info()
Out[ ]:
    <class 'pandas.core.frame.DataFrame'>
    DatetimeIndex: 4734 entries, 2000-01-03 to 2018-10-24
    Data columns (total 2 columns):
    SPX    4734 non-null float64
    VIX    4734 non-null float64
    dtypes: float64(2)
    memory usage: 111.0 KB
```

Let's plot the S&P 500 and VIX to see how they look from the year 2010 onwards:

```
In [ ]:
    %matplotlib inline
    import matplotlib.pyplot as plt

    plt.figure(figsize = (12, 8))

    ax_spx = df['SPX'].plot()
    ax_vix = df['VIX'].plot(secondary_y=True)

    ax_spx.legend(loc=1)
    ax_vix.legend(loc=2)

    plt.show();
```

This gives us the following graph:

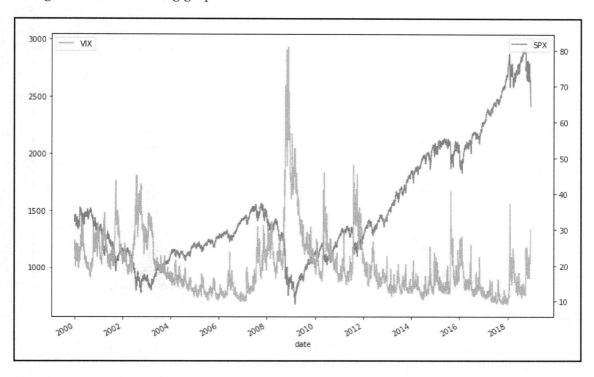

Observe that when the S&P 500 moves up, the VIX appears to move downwards, exhibiting a negative correlation relationship. We need to perform more statistical analysis to be sure.

Perhaps we might be interested in the daily returns of both the indexes. The `diff()` method returns the set of differences between the prior period values. A histogram can be used to give us a rough sense of the data density estimation over a bin interval of 100:

```
In [ ]:
    df.diff().hist(
        figsize=(10, 5),
        color='blue',
        bins=100);
```

The `hist()` method gives us the following histograms:

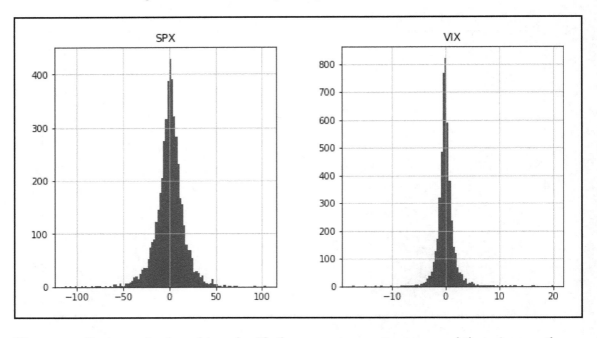

The same effect can also be achieved with the `pct_change()` command that gives us the percentage change over the prior period values:

```
In [ ]:
    df.pct_change().hist(
        figsize=(10, 5),
        color='blue',
        bins=100);
```

We get the same histogram in terms of percentage changes:

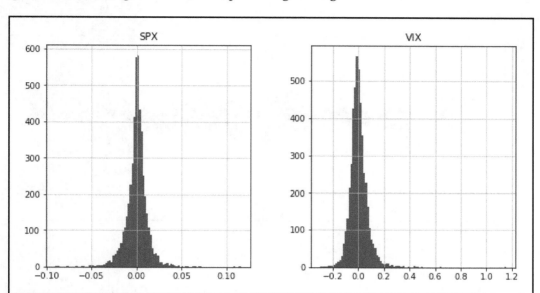

For quantitative analysis of returns, we are interested in the logarithm of daily returns. Why use log returns over simple returns? There are several reasons, but the most important of them is normalization, and this avoids the problem of negative prices.

We can use the `shift()` function of `pandas` to shift the values by a certain number of periods. The `dropna()` method removes the unused values at the end of the logarithmic calculation transformation. The `log()` method of NumPy helps to calculate the logarithm of all values in the DataFrame object as a vector, and will be stored in the `log_returns` variable as a DataFrame object. The logarithm values can then be plotted to give us a graph of daily log returns. Here is the code to plot the logarithm values:

```
In [ ]:
    import numpy as np

    log_returns = np.log(df / df.shift(1)).dropna()
    log_returns.plot(
        subplots=True,
        figsize=(10, 8),
        color='blue',
        grid=True
    );
    for ax in plt.gcf().axes:
        ax.legend(loc='upper left')
```

We get the following output:

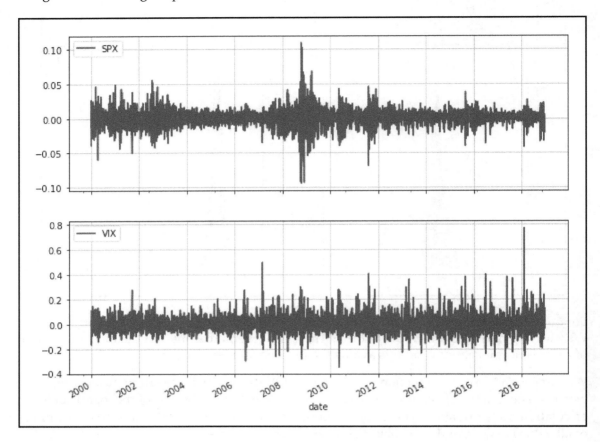

The top and bottom graph show the log returns of SPX and VIX, respectively, from the year 2000 and up to the present period.

The correlation between the SPX and the VIX

We can use the `corr()` method to derive the correlation values between each column of values in the `pandas` DataFrame object, as in the following Python example:

```
In [ ]:
    log_returns.corr()
```

This gives us the following correlation table:

	SPX	VIX
SPX	1.000000	-0.733161
VIX	-0.733161	1.000000

At -0.731433, the SPX is negatively correlated with the VIX. To help us better visualize this relationship, we can plot both sets of the daily log return values as a scatter plot. The `statsmodels.api` module is used to obtain the ordinary least squares regression line between the scattered data:

```
In [ ]:
    import statsmodels.api as sm

    log_returns.plot(
        figsize=(10,8),
        x="SPX",
        y="VIX",
        kind='scatter')

    ols_fit = sm.OLS(log_returns['VIX'].values,
    log_returns['SPX'].values).fit()

    plt.plot(log_returns['SPX'], ols_fit.fittedvalues, 'r');
```

We get the following output:

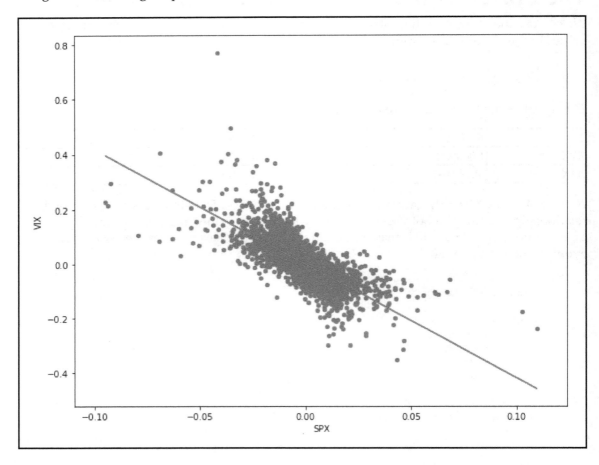

The downward-sloping regression line, as shown in the preceding graph, confirms the negative correlation relationship between the S&P 500 and the VIX indices.

The `rolling().corr()` method of `pandas` computes the moving-window correlation between two time series. We use a value of `252` to represent the number of trading days in the moving window to compute the annual rolling correlation, using the following commands:

```
In [ ]:
    plt.ylabel('Rolling Annual Correlation')

    df_corr = df['SPX'].rolling(252).corr(other=df['VIX'])
    df_corr.plot(figsize=(12,8));
```

We get the following output:

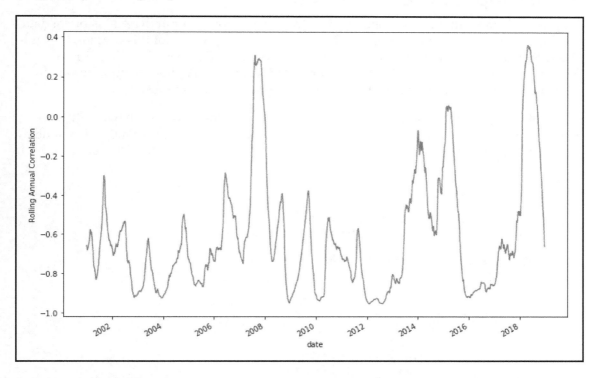

It can be seen from the preceding graph that the SPX and the VIX are negatively correlated, fluctuating between 0.0 and -0.9 during most of the lifetime of the indices using 252 trading days per year.

Calculating the VIX Index

In this section, we will perform a step-by-step replication of the VIX Index. Calculation of the VIX Index is documented on the CBOE website. You may obtain a copy of the CBOE VIX white paper at `http://www.cboe.com/micro/vix/vixwhite.pdf`.

Importing SPX options data

Suppose you had gathered SPX options data from your broker or purchased historical data from an external source such as the CBOE website. For the purpose of this chapter, the end-of-day SPX option chain prices are observed from Monday, October 15, 2018 to Friday, October 19, 2018 and saved into **Comma-separated Values** (**CSV**) files. Sample copies of these files are provided under the **files** folder of the source codes repository.

In the following example, write a function, `read_file()`, that accepts the file path in its first argument indicating the location of a CSV file, and returns a tuple of a list of metadata and a list of option chain data:

```
In [ ]:
    import csv

    META_DATA_ROWS = 3  # Header data starts at line 4
    COLS = 7  # Each option data occupy 7 columns

    def read_file(filepath):
        meta_rows = []
        calls_and_puts = []

        with open(filepath, 'r') as file:
            reader = csv.reader(file)
            for row, cells in enumerate(reader):
                if row < META_DATA_ROWS:
                    meta_rows.append(cells)
                else:
                    call = cells[:COLS]
                    put = cells[COLS:-1]

                    calls_and_puts.append((call, put))

        return (meta_rows, calls_and_puts)
```

Note that the structure of your own options data may differ from this example. Be prudent to check and modify this function accordingly. Having imported our dataset, we can proceed to parse and extract useful information.

Parsing SPX options data

In this example, we assume the top three rows of the CSV file contain meta information, with the rest of the options chain prices starting from row four onwards. For each row of options pricing data, the first seven columns contain the bid and ask quotes of a call contract, with the next seven columns for a put contract. The first of each seven column contains a string describing the expiry date, strike price, and contract code. Follow these steps to parse information from our CSV file:

1. Each row of meta information is appended in a list variable named `meta_data`, while each row of options data is appended to a list variable named `calls_and_puts`. Reading a single file with this function gives us the following:

    ```
    In [ ]:
        (meta_rows, calls_and_puts) = \
            read_file('files/chapter07/SPX_EOD_2018_10_15.csv')
    ```

2. Printing each row of metadata provides the following:

    ```
    In [ ]:
        for line in meta_rows:
            print(line)
    Out[ ]:
        ['SPX (S&P 500 INDEX)', '2750.79', '-16.34']
        ['Oct 15 2018 @ 20:00 ET']
        ['Calls', 'Last Sale', 'Net', 'Bid', 'Ask', 'Vol', 'Open Int',
    'Puts', 'Last Sale', 'Net', 'Bid', 'Ask', 'Vol', 'Open Int']
    ```

3. The current time of the option quotes can be found on the second row of our metadata. Since Eastern Time is 5 hours behind **Greenwich Mean Time (GMT)**, we replace the `ET` string and parse the entire string as a `datetime` object. The following function, `get_dt_current()`, demonstrates this:

    ```
    In [ ]:
        from dateutil import parser

        def get_dt_current(meta_rows):
            """
            Extracts time information.

            :param meta_rows: 2D array
            :return: parsed datetime object
            """
            # First cell of second row contains time info
            date_time_row = meta_rows[1][0]
    ```

```
# Format text as ET time string
current_time = date_time_row.strip()\
    .replace('@ ', '')\
    .replace('ET', '-05:00')\
    .replace(',', '')

dt_current =  parser.parse(current_time)
return dt_current
```

4. From the meta information of our options data, extract the date and time information as Chicago local time:

```
In [ ]:
    dt_current =  get_dt_current(meta_rows)
    print(dt_current)
Out[ ]:
    2018-10-15 20:00:00-05:00
```

5. Now, let's look at the first two lines of our option quotes data:

```
In [ ]:
    for line in calls_and_puts[:2]:
        print(line)
Out[ ]:
    (['2018 Oct 15 1700.00 (SPXW1815J1700)', '0.0', '0.0',
'1039.30', '1063.00', '0',      '0'], ['2018 Oct     15 1700.00
(SPXW1815V1700)', '0.15', '0.0', ' ', '0.05', '0'])
    (['2018 Oct 15 1800.00 (SPXW1815J1800)', '0.0', '0.0',
'939.40', '963.00', '0',      '0'], ['2018 Oct     15 1800.00
(SPXW1815V1800)', '0.10', '0.0', ' ', '0.05', '0'])
```

Each item in the list contains a tuple of two objects, with each object containing a list of a call option and a put option pricing data having the same strike price. Referring to our printed headers, the seven items of each option price list data contains the contract code with expiry date, the last sale price, the net change in price, the bid price, ask price, volume traded, and open interest.

Let's write a function to parse the description of each SPX option dataset:

```
In [ ]:
    from decimal import Decimal

    def parse_expiry_and_strike(text):
        """
        Extracts information about the contract data.

        :param text: the string to parse.
```

```
        :return: a tuple of expiry date and strike price
        """
        # SPXW should expire at 3PM Chicago time.
        [year, month, day, strike, option_code] = text.split(' ')
        expiry = '%s %s %s 3:00PM -05:00' % (year, month, day)
        dt_object = parser.parse(expiry)

        """
        Third friday SPX standard options expire at start of trading
        8.30 A.M. Chicago time.
        """
        if is_third_friday(dt_object):
            dt_object = dt_object.replace(hour=8, minute=30)

        strike = Decimal(strike)
        return (dt_object, strike)
```

The utility function, `parse_expiry_and_strike()`, returns a tuple of expiration date object, and the strike price as a `Decimal` object.

Each contract data is a string containing the expiration year, month, day, and strike price, followed by the contract code, all separated by spaces. We take the date components and reconstruct a date and time string, easily parsed by the parser function of `dateutil` imported earlier. Weekly options expire at 4 PM New York time, or 3 PM Chicago time. Standard third Friday options are AM-settled and expire at the start of the trading day at 8.30 AM. We replace the expiration time in accordance with performing a `is_third_friday()` check, implemented as follows:

```
In [ ]:
    def is_third_friday(dt_object):
        return dt_object.weekday() == 4 and 15 <= dt_object.day <= 21
```

Testing our function with a simple contract code data and printing the results gives us the following:

```
In [ ]:
    test_contract_code = '2018 Sep 26 1800.00 (*)'
    (expiry, strike) = parse_expiry_and_strike(test_contract_code)
In [ ]:
    print('Expiry:', expiry)
    print('Strike price:', strike)
Out[ ]:
    Expiry: 2018-09-26 15:00:00-05:00
    Strike price: 1800.00
```

Since September 26, 2018 falls on a Wednesday, the SPXW option will expire at 3 PM CDT, Chicago local time.

This time round, let's test our function with a contract code data that falls on a third Friday:

```
In [ ]:
    test_contract_code = '2018 Oct 19 2555.00 (*)'
    (expiry, strike) = parse_expiry_and_strike(test_contract_code)
In [ ]:
    print('Expiry:', expiry)
    print('Strike price:', strike)
Out[ ]:
    Expiry: 2018-10-19 08:30:00-05:00
    Strike price: 2555.00
```

The test contract code data we have used is October 19, 2018, which falls on the third Friday of October. This is a standard SPX option that is settled at the start of the trading day, at 8.30 AM Chicago time.

With our utility function in place, we can now go ahead and parse a single call or put option price entry, and return useful information that we can use:

```
In [ ]:
    def format_option_data(option_data):
        [desc, _, _, bid_str, ask_str] = option_data[:5]
        bid = Decimal(bid_str.strip() or '0')
        ask = Decimal(ask_str.strip() or '0')
        mid = (bid+ask) / Decimal(2)
        (expiry, strike) = parse_expiry_and_strike(desc)
        return (expiry, strike, bid, ask, mid)
```

The utility function, `format_option_data()`, takes `option_data` as its argument, containing a list of data we have seen earlier. The descriptive data at index zero contains the contract code data that we can parse using the `parse_expiry_and_strike()` function. Indexes three and four contain the bid and ask price, which are used to compute the mid price. The mid price is the average of the bid and ask price. This function returns a tuple of the option expiry date, as well as the strike, bid, ask, and mid prices as `Decimal` objects.

Finding near-term and next-term options

The VIX Index measures the 30-day expected volatility of the SPX using the market quotes of call and put options expiring between 24 days and 36 days. In between these dates, there will be two SPX option contract maturity dates. Options expiring the nearest are referred to as near-term options, while the options expiring later are referred to as next-term options. Happening once per week, when the option maturities are out of the 24 to 36 days range, new contract maturities will be chosen as the new near-term and next-term options.

To help us in finding the near-term and next-term options, let's organize our puts and calls options data indexed by maturity dates, each having a `pandas` DataFrame indexed by strike price. We would need the following DataFrame column definitions:

```
In [ ]:
    CALL_COLS = ['call_bid', 'call_ask', 'call_mid']
    PUT_COLS = ['put_bid', 'put_ask', 'put_mid']
    COLUMNS = CALL_COLS + PUT_COLS + ['diff']
```

The following function, `generate_options_chain()`, organizes our list dataset, `calls_and_puts`, into a single dictionary variable chain:

```
In [ ]:
    import pandas as pd

    def generate_options_chain(calls_and_puts):
        chain = {}

        for row in calls_and_puts:
            (call, put) = row

            (call_expiry, call_strike, call_bid, call_ask, call_mid) = \
                format_option_data(call)
            (put_expiry, put_strike, put_bid, put_ask, put_mid) = \
                format_option_data(put)

            # Ensure each line contains the same put and call maturity
            assert(call_expiry == put_expiry)

            # Get or create the DataFrame at the expiry
            df = chain.get(call_expiry, pd.DataFrame(columns=COLUMNS))

            df.loc[call_strike, CALL_COLS] = \
                [call_bid, call_ask, call_mid]
            df.loc[call_strike, PUT_COLS] = \
                [put_bid, put_ask, put_mid]
            df.loc[call_strike, 'diff'] = abs(put_mid-call_mid)

            chain[call_expiry] = df

        return chain
In [ ]:
    chain = generate_options_chain(calls_and_puts)
```

The `chain` variable has its keys as the option maturity dates, each referencing to a `pandas` DataFrame object. Two invocations to the `format_option_data()` function are made to derive the calls and puts data of interest. The `assert` keyword ensures the integrity of our call and put expiry dates, based on the assumption that each line in our dataset refers to the same expiry date. Otherwise, an exception will be thrown and calls to our attention to inspect the dataset for any signs of corruption.

The `loc` keyword assigns column values for a specific strike price, for both calls and puts data. In addition, the `diff` column contains the absolute difference between the mid prices of call and put quotes, which we will use later.

Let's view the first and last two keys in our `chain` dictionary:

```
In [ ]:
    chain_keys = list(chain.keys())
    for row in chain_keys[:2]:
        print(row)
    print('...')
    for row in chain_keys[-2:]:
        print(row)
Out[ ]:
    2018-10-15 15:00:00-05:00
    2018-10-17 15:00:00-05:00
    ...
    2020-06-19 08:30:00-05:00
    2020-12-18 08:30:00-05:00
```

Our dataset contains options prices for maturities over the next two years. From these, we select our near-term and next-term maturities with the following function:

```
In [ ]:
    def find_option_terms(chain, dt_current):
        """
        Find the near-term and next-term dates from
        the given indexes of the dictionary.

        :param chain: dictionary object
        :param dt_current: DateTime object of option quotes
        :return: tuple of 2 datetime objects
        """
        dt_near = None
        dt_next = None

        for dt_object in chain.keys():
            delta = dt_object - dt_current
            if delta.days > 23:
```

```
            # Skip non-fridays
            if dt_object.weekday() != 4:
                continue

            # Save the near term date
            if dt_near is None:
                dt_near = dt_object
                continue

            # Save the next term date
            if dt_next is None:
                dt_next = dt_object
                break

        return (dt_near, dt_next)
Out[ ]:
    (dt_near, dt_next) = find_option_terms(chain, dt_current)
```

Here, we are simply selecting the first two options whose maturities are over 23 days from the time of this dataset. These two option maturities are as follows:

```
In [ ]:
    print('Found near-term maturity', dt_near,
          'with', dt_near-dt_current, 'to expiry')
    print('Found next-term maturity', dt_next,
          'with', dt_next-dt_current, 'to expiry')
Out[ ]:
    Found near-term maturity 2018-11-09 15:00:00-05:00 with 24 days,
19:00:00 to expiry
    Found next-term maturity 2018-11-16 08:30:00-05:00 with 31 days,
12:30:00 to expiry
```

The near-term maturity is on November 9, 2018 and the next-term maturity is on November 16, 2018.

Calculating the required minutes

The formula for calculating the VIX is given as follows:

$$VIX = 100 \times \sqrt{\left\{T_1\sigma_1^2\left[\frac{N_{T_2} - N_{30}}{N_{T_2} - N_{T_1}}\right] + T_2\sigma_2^2\left[\frac{N_{30} - N_{T_1}}{N_{T_2} - N_{T_1}}\right]\right\} \times \frac{N_{365}}{N_{30}}}$$

Here, the following applies:

- T_1 is the number of years to settlement of the near-term options
- T_2 is the number of years to settlement of the next-term options
- N_{T1} is the number of minutes to settlement of the near-term options
- N_{T2} is the number of minutes to settlement of the next-term options
- N_{30} is the number of minutes in 30 days
- N_{365} is the number of minutes in a year with 365 days

Let's find out these values in Python:

```
In [ ]:
    dt_start_year = dt_current.replace(
        month=1, day=1, hour=0, minute=0, second=0)
    dt_end_year = dt_start_year.replace(year=dt_current.year+1)

    N_t1 = Decimal((dt_near-dt_current).total_seconds() // 60)
    N_t2 = Decimal((dt_next-dt_current).total_seconds() // 60)
    N_30 = Decimal(30 * 24 * 60)
    N_365 = Decimal((dt_end_year-dt_start_year).total_seconds() // 60)
```

The difference of two `datetime` objects returns a `timedelta` object, whose `total_seconds()` method gives the difference in terms of the number of seconds. The number of minutes can be obtained by dividing the number of seconds by sixty. The number of minutes in a year is found by taking the difference between the start of next year and the start of current year, while the number of minutes in a month is simply the sum of seconds in thirty days.

The values obtained are as follows:

```
In [ ]:
    print('N_365:', N_365)
    print('N_30:', N_30)
    print('N_t1:', N_t1)
    print('N_t2:', N_t2)
Out[ ]:
    N_365: 525600
    N_30: 43200
    N_t1: 35700
    N_t2: 45390
```

The general formula for calculating T is as follows:

$$T = \{M_{current\ day} + M_{other\ days} + M_{settlement\ day}\}/\text{minutes in a year}$$

Here, the following applies:

- $M_{current\ day}$ is the number of minutes remaining until midnight of the current day
- $M_{other\ days}$ is the sum of minutes between the current day and the expiration day
- $M_{settlement\ day}$ is the number of minutes from midnight of the expiration day until the expiration time

With these, we can find T_1 and T_2, that is, the amount of time remaining per year for the near-term and next-term options:

```
In [ ]:
    t1 = N_t1 / N_365
    t2 = N_t2 / N_365
In [ ]:
    print('t1:%.5f'%t1)
    print('t2:%.5f'%t2)
Out [ ]:
    t1:0.06792
    t2:0.08636
```

The near-term option is 0.6792 years to maturity and the next-term option is 0.08636 years to maturity.

Calculating the forward SPX Index level

For each contract month, the forward SPX level F is given as follows:

$$F = \text{Strike price} + e^{rT}(\text{call price} - \text{put price})$$

Here, the strike price is chosen where the absolute difference between the call and put prices is the minimum. Note that options with zero bid prices are not taken into account for the VIX Index calculation. This suggests that as volatility of the SPX and options changes, the bid quotes may become zero, and the number of options used in calculation of the VIX Index may vary at any minute!

We can represent the forward index level calculation with the `determine_forward_level()` function, as shown in the following code:

```
In [ ]:
    import math

    def determine_forward_level(df, r, t):
```

```
"""
Calculate the forward SPX Index level.

:param df: pandas DataFrame for a single option chain
:param r: risk-free interest rate for t
:param t: time to settlement in years
:return: Decimal object
"""
min_diff = min(df['diff'])
pd_k = df[df['diff'] == min_diff]
k = pd_k.index.values[0]

call_price = pd_k.loc[k, 'call_mid']
put_price = pd_k.loc[k, 'put_mid']
return k + Decimal(math.exp(r*t))*(call_price-put_price
```

The `df` argument is the DataFrame containing the near-term or next-term options prices.
The `min_diff` variable contains the minimum of all absolute price differences computed in
the diff column earlier. The `pd_k` variable contains the DataFrame at which we will choose
our strike price having the minimum absolute price difference.

Note that we are assuming an interest rate of 2.17% for both option chains for the sake of
simplicity. In practice, the interest rates of near-term and next-term options are based on a
cubic spline calculation of U.S. Treasury yield curve rates, or **Constant Maturity Treasury
rates (CMTs)**. Yield curve rates are available from the U.S. Department of the Treasury
website at `https://www.treasury.gov/resource-center/data-chart-center/interest-rates/Pages/TextView.aspx?data=yieldYearyear=2018`.

Let's calculate the forward SPX level for the near-term options as `f1`:

```
In [ ]:
    r = Decimal(2.17/100)
In [ ]:
    df_near = chain.get(dt_near)
    f1 = determine_forward_level(df_near, r, t1)
In [ ]:
    print('f1:', f1)
Out[ ]:
    f1: 2747.596459994546094129930225
```

We will be using the forward SPX level, *F*, as 2747.596.

Finding the required forward strike prices

The forward strike price is the strike price immediately below the forward SPX level, denoted by k0, and is determined by the find_k0() function, written as follows:

```
In [ ]:
    def find_k0(df, f):
        return df[df.index<f].tail(1).index.values[0]
```

The value of k0 of the near-term option can simply be found with the function call:

```
In [ ]:
    k0_near = find_k0(df_near, f1)
In [ ]:
    print('k0_near:', k0_near)
Out[ ]:
    k0_near: 2745.00
```

The near-term forward strike price is found to be 2745.

Determining strike price boundaries

When selecting options used in the VIX Index calculation, calls and puts with bid prices of zero are ignored. For far **out-of-the-money** (**OTM**) put options where strike prices are lower than k0, the lower price boundary terminates when two consecutive zero bid prices are encountered. Similarly, for far OTM call options with strike prices more than k0, the upper price boundary terminates when two consecutive zero bid prices are encountered.

The following function, find_lower_and_upper_bounds(), illustrates the process of finding the lower and upper boundaries in Python code:

```
In [ ]:
    def find_lower_and_upper_bounds(df, k0):
        """
        Find the lower and upper boundary strike prices.

        :param df: the pandas DataFrame of option chain
        :param k0: the forward strike price
        :return: a tuple of two Decimal objects
        """
        # Find lower bound
        otm_puts = df[df.index<k0].filter(['put_bid', 'put_ask'])
        k_lower = 0
        for i, k in enumerate(otm_puts.index[::-1][:-2]):
            k_lower = k
```

```
        put_bid_t1 = otm_puts.iloc[-i-1-1]['put_bid']
        put_bid_t2 = otm_puts.iloc[-i-1-2]['put_bid']
        if put_bid_t1 == 0 and put_bid_t2 == 0:
            break
        if put_bid_t2 == 0:
            k_lower = otm_puts.index[-i-1-1]

    # Find upper bound
    otm_calls = df[df.index>k0].filter(['call_bid', 'call_ask'])
    k_upper = 0
    for i, k in enumerate(otm_calls.index[:-2]):
        call_bid_t1 = otm_calls.iloc[i+1]['call_bid']
        call_bid_t2 = otm_calls.iloc[i+2]['call_bid']
        if call_bid_t1 == 0 and call_bid_t2 == 0:
            k_upper = k
            break

    return (k_lower, k_upper)
```

The df argument is the pandas DataFrame of the option prices. The otm_puts variable contains OTM puts data, and is iterated backwards in descending order by the for loop. At each iteration, the k_lower variable stores the current strike price while we are looking at two bid prices ahead in the loop. When the for loop terminates due to two zero bids encountered or reaches the end of the list, k_lower will contain the strike price of the lower boundary.

The same methodology is applied in finding the strike price of the upper boundary. Since the strike prices of OTM calls are already in descending order, we simply read the prices using forward index referencing on the iloc command.

When we provide the near-term option chain data to this function, the lower and upper strike price boundaries can be obtained from the k_lower and k_upper variables, respectively, as shown in the following codes:

```
In [ ]:
    (k_lower_near, k_upper_near) = \
        find_lower_and_upper_bounds(df_near, k0_near)
In [ ]:
    print(k_lower_near, k_upper_near)
Out[ ]:
    1250.00 3040.00
```

Near-term options with strike prices from 1,500 to 3,200 will be used in the calculation of the VIX Index.

Tabulating contributions by strike prices

Since the VIX Index is composed of prices of calls and puts options expiring on an average of 30 days, each option of the chosen maturity date contributes to the VIX Index calculation by a certain amount. This amount is given as the general formula:

$$\frac{\triangle K_i}{K_i^2} e^{RT} \,(\text{midpoint of bid-ask spread at } K_i)$$

Here, T is the time to expiration of the option, R is the risk-free interest rate to expiration of the option, K_i is the strike price of the ith OTM option, and $\triangle K_i$ is the half-difference on either side of K_i such that $\triangle K_i = 0.5(K_{i+1}-K_{i-1})$.

We can represent this formula by the following `calculate_contrib_by_strike()` function:

```
In [ ]:
    def calculate_contrib_by_strike(delta_k, k, r, t, q):
        return (delta_k / k**2)*Decimal(math.exp(r*t))*q
```

In calculating $\triangle K_i = 0.5(K_{i+1}-K_{i-1})$, we search for K_{i-1} using the utility function, `find_prev_k()`, as follows:

```
In [ ]:
    def find_prev_k(k, i, k_lower, df, bid_column):
        """
        Finds the strike price immediately below k
        with non-zero bid.

        :param k: current strike price at i
        :param i: current index of df
        :param k_lower: lower strike price boundary of df
        :param bid_column: The column name that reads the bid price.
            Can be 'put_bid' or 'call_bid'.
        :return: strike price as Decimal object.
        """
        if k <= k_lower:
            k_prev = df.index[i-1]
            return k_prev

        # Iterate backwards to find put bids
        k_prev = 0
        prev_bid = 0
        steps = 1
        while prev_bid == 0:
```

```
            k_prev = df.index[i-steps]
            prev_bid = df.loc[k_prev][bid_column]
            steps += 1

        return k_prev
```

Similarly, we use the same procedure in searching for K_{i+1} using the utility function, find_next_k(), as follows:

```
In [ ]:
    def find_next_k(k, i, k_upper, df, bid_column):
        """
        Finds the strike price immediately above k
        with non-zero bid.

        :param k: current strike price at i
        :param i: current index of df
        :param k_upper: upper strike price boundary of df
        :param bid_column: The column name that reads the bid price.
            Can be 'put_bid' or 'call_bid'.
        :return: strike price as Decimal object.
        """
        if k >= k_upper:
            k_next = df.index[i+1]
            return k_next

        k_next = 0
        next_bid = 0
        steps = 1
        while next_bid == 0:
            k_next = df.index[i+steps]
            next_bid = df.loc[k_next][bid_column]
            steps += 1

        return k_next
```

With the preceding utility functions, we can now create a function, tabulate_contrib_by_strike(), that uses an iterative procedure to calculate the contributions of each option for every strike price available in the pandas DataFrame df with option prices, returning a new DataFrame containing the final dataset used towards the calculation of the VIX Index:

```
In [ ]:
    import pandas as pd

    def tabulate_contrib_by_strike(df, k0, k_lower, k_upper, r, t):
        """
```

```
Computes the contribution to the VIX index
for every strike price in df.

:param df: pandas DataFrame containing the option dataset
:param k0: forward strike price index level
:param k_lower: lower boundary strike price
:param k_upper: upper boundary strike price
:param r: the risk-free interest rate
:param t: the time to expiry, in years
:return: new pandas DataFrame with contributions by strike price
"""
COLUMNS = ['Option Type', 'mid', 'contrib']
pd_contrib = pd.DataFrame(columns=COLUMNS)

for i, k in enumerate(df.index):
    mid, bid, bid_column = 0, 0, ''
    if k_lower <= k < k0:
        option_type = 'Put'
        bid_column = 'put_bid'
        mid = df.loc[k]['put_mid']
        bid = df.loc[k][bid_column]
    elif k == k0:
        option_type = 'atm'
    elif k0 < k <= k_upper:
        option_type = 'Call'
        bid_column = 'call_bid'
        mid = df.loc[k]['call_mid']
        bid = df.loc[k][bid_column]
    else:
        continue  # skip out-of-range strike prices

    if bid == 0:
        continue  # skip zero bids

    k_prev = find_prev_k(k, i, k_lower, df, bid_column)
    k_next = find_next_k(k, i, k_upper, df, bid_column)
    delta_k = Decimal((k_next-k_prev)/2)

    contrib = calculate_contrib_by_strike(delta_k, k, r, t, mid)
    pd_contrib.loc[k, COLUMNS] = [option_type, mid, contrib]

return pd_contrib
```

The resulting DataFrame is indexed by strike price and contains three columns—the option type as either a *Call* or *Put,* the average of the bid-ask spread, and the contribution to the VIX Index.

Tabulating the contributions of our near-term options gives us the following:

```
In [ ]:
    pd_contrib_near = tabulate_contrib_by_strike(
        df_near, k0_near, k_lower_near, k_upper_near, r, t1)
```

Peeking at the head of the results provides the following:

```
In [ ]:
    pd_contrib_near.head()
```

This gives us the following table:

	Option Type	mid	contrib
1250.00	Put	0.10	0.00000320472000727187449342636826
1300.00	Put	0.125	0.00000370367974213188157986590101
1350.00	Put	0.15	0.00000412129630564798674566147997
1400.00	Put	0.20	0.00000510956633812479989385581445
1450.00	Put	0.20	0.00000476325803696770881900470693

Peeking at the tail of the results provides the following:

```
In [ ]:
    pd_contrib_near.tail()
```

This also gives us the following table:

	Option Type	mid	contrib
3020.00	Call	0.175	9.60802845257229048941134356E-8
3025.00	Call	0.225	1.231237623174939828257858985E-7
3030.00	Call	0.175	9.54471377521161522068938969E-8
3035.00	Call	0.20	1.08723324234557377460190108E-7
3040.00	Call	0.15	8.12744818759030454030476026E-8

The `pd_contrib_near` variable contains the near-term call and put OTM options contained in a single DataFrame.

Calculating the volatilities

The volatility calculation for the chosen options is given as follows:

$$\sigma^2 = \frac{2}{T}\sum_i \frac{\triangle K_i}{K_i^2}e^{RT}(\text{midpoint of bid-ask spread at } K_i) - \frac{1}{T}\left[\frac{F}{K_0} - 1\right]^2$$

Since we have already computed the contributions for the summation term, this formula can be simply written in Python as the `calculate_volatility()` function:

```
In [ ]:
    def calculate_volatility(pd_contrib, t, f, k0):
        """
        Calculate the volatility for a single-term option

        :param pd_contrib: pandas DataFrame
            containing contributions by strike
        :param t: time to settlement of the option
        :param f: forward index level
        :param k0: immediate strike price below the forward level
        :return: volatility as Decimal object
        """
        term_1 = Decimal(2/t)*pd_contrib['contrib'].sum()
        term_2 = Decimal(1/t)*(f/k0 - 1)**2
        return term_1 - term_2
```

Computing the volatility of the near-term option gives us the following:

```
In [ ]:
    volatility_near = calculate_volatility(
        pd_contrib_near, t1, f1, k0_near)
In [ ]:
    print('volatility_near:', volatility_near)
Out[ ]:
    volatility_near: 0.04891704334249740486501736967
```

The volatility of the near-term option is 0.04891.

Calculating the next-term options

Just as we did for near-term options, calculating the next-term options is pretty straightforward with the following Python calls to the functions already defined in place:

```
In [ ] :
    df_next = chain.get(dt_next)

    f2 = determine_forward_level(df_next, r, t2)
    k0_next = find_k0(df_next, f2)
    (k_lower_next, k_upper_next) = \
        find_lower_and_upper_bounds(df_next, k0_next)
    pd_contrib_next = tabulate_contrib_by_strike(
        df_next, k0_next, k_lower_next, k_upper_next, r, t2)
    volatility_next = calculate_volatility(
        pd_contrib_next, t2, f2, k0_next)
In [ ]:
    print('volatility_next:', volatility_next)
Out[ ]:
    volatility_next: 0.04524308316212813982254693873
```

Since `dt_next` is our next-term maturity date, calling `chain.get()` retrieves the next-term option prices from the options chain store. With this data, we determine the forward SPX level, `f2`, for the next-term option, find its forward strike price, `k0_next`, and find its lower and upper strike price boundaries. Next, we tabulate the contributions of each option in calculating the VIX Index within the boundary of strike prices, from which we calculate the next-term volatility with the `calculate_volatility()` function.

The volatility of the next-term option is 0.0452.

Calculating the VIX Index

Finally, the 30-day weighted average of the VIX Index is written as follows:

$$VIX = 100 \times \sqrt{\left\{T_1\sigma_1^2\left[\frac{N_{T_2} - N_{30}}{N_{T_2} - N_{T_1}}\right] + T_2\sigma_2^2\left[\frac{N_{30} - N_{T_1}}{N_{T_2} - N_{T_1}}\right]\right\} \times \frac{N_{365}}{N_{30}}}$$

Representing this formula in Python code gives us the following:

```
In [ ]:
    def calculate_vix_index(t1, volatility_1, t2,
                            volatility_2, N_t1, N_t2, N_30, N_365):
        inner_term_1 = t1*Decimal(volatility_1)*(N_t2-N_30)/(N_t2-N_t1)
        inner_term_2 = t2*Decimal(volatility_2)*(N_30-N_t1)/(N_t2-N_t1)
        sqrt_terms = math.sqrt((inner_term_1+inner_term_2)*N_365/N_30)
        return 100 * sqrt_terms
```

Substituting with the values from near-term and next-term options provides the following:

```
In [ ]:
    vix = calculate_vix_index(
        t1, volatility_near, t2,
        volatility_next, N_t1, N_t2,
        N_30, N_365)
In [ ]:
    print('At', dt_current, 'the VIX is', vix)
Out[ ]:
    At 2018-10-15 20:00:00-05:00 the VIX is 21.431114075693934
```

We obtained the VIX Index as 21.43 at the close of October 15, 2018.

Calculating multiple VIX indexes

With a single VIX value calculated for a particular trading day, we can reuse the defined functions to calculate VIX values over a time period.

Let's write a function, `process_file()`, to process a single file path, and return the calculated VIX Index:

```
In [ ]:
    def process_file(filepath):
        """
        Reads the filepath and calculates the VIX index.

        :param filepath: path the options chain file
        :return: VIX index value
        """
        headers, calls_and_puts = read_file(filepath)
        dt_current = get_dt_current(headers)

        chain = generate_options_chain(calls_and_puts)
        (dt_near, dt_next) = find_option_terms(chain, dt_current)
```

```
N_t1 = Decimal((dt_near-dt_current).total_seconds() // 60)
N_t2 = Decimal((dt_next-dt_current).total_seconds() // 60)
t1 = N_t1 / N_365
t2 = N_t2 / N_365

# Process near-term options
df_near = chain.get(dt_near)
f1 = determine_forward_level(df_near, r, t1)
k0_near = find_k0(df_near, f1)
(k_lower_near, k_upper_near) = find_lower_and_upper_bounds(
    df_near, k0_near)
pd_contrib_near = tabulate_contrib_by_strike(
    df_near, k0_near, k_lower_near, k_upper_near, r, t1)
volatility_near = calculate_volatility(
    pd_contrib_near, t1, f1, k0_near)

# Process next-term options
df_next = chain.get(dt_next)
f2 = determine_forward_level(df_next, r, t2)
k0_next = find_k0(df_next, f2)
(k_lower_next, k_upper_next) = find_lower_and_upper_bounds(
    df_next, k0_next)
pd_contrib_next = tabulate_contrib_by_strike(
    df_next, k0_next, k_lower_next, k_upper_next, r, t2)
volatility_next = calculate_volatility(
    pd_contrib_next, t2, f2, k0_next)

vix = calculate_vix_index(
    t1, volatility_near, t2,
    volatility_next, N_t1, N_t2,
    N_30, N_365)

return vix
```

Suppose we have observed options chain data and collected it into CSV files for the week of 15th to 19th of October, 2018. We can define the filenames and file path pattern into constant variables:

```
In [ ]:
    FILE_DATES = [
        '2018_10_15',
        '2018_10_16',
        '2018_10_17',
        '2018_10_18',
        '2018_10_19',
    ]
    FILE_PATH_PATTERN = 'files/chapter07/SPX_EOD_%s.csv'
```

Iterating through the dates and setting the calculated VIX value into a `pandas` DataFrame with a column named `'VIX'` gives us the following:

```
In [ ] :
    pd_calcs = pd.DataFrame(columns=['VIX'])

    for file_date in FILE_DATES:
        filepath = FILE_PATH_PATTERN % file_date

        vix = process_file(filepath)
        date_obj = parser.parse(file_date.replace('_', '-'))

        pd_calcs.loc[date_obj, 'VIX'] = vix
```

Observing our data with the `head()` command provides the following:

```
In [ ]:
    pd_calcs.head(5)
```

This gives us the following table, containing VIX values over a 5-day period:

	VIX
2018-10-15	21.4311
2018-10-16	17.7384
2018-10-17	17.4741
2018-10-18	20.0477
2018-10-19	19.9196

Comparing the results

Let's compare the calculated VIX values against actual VIX values by reusing the DataFrame `df_vix_data` VIX Index that was downloaded in an earlier section, and extract only the relevant values for the corresponding week of 15th to 19th of October, 2018:

```
In [ ]:
    df_vix = df_vix_data['2018-10-14':'2018-10-21']['5. adjusted close']
```

The actual end-of-day VIX values for the period are as follows:

```
In [ ]:
    df_vix.head(5)
Out [ ]:
    date
    2018-10-15    21.30
    2018-10-16    17.62
```

```
2018-10-17     17.40
2018-10-18     20.06
2018-10-19     19.89
Name: 5. adjusted close, dtype: float64
```

Let's merge the actual VIX values and the calculated values into a single DataFrame, and plot them:

```
In [ ]:
    df_merged = pd.DataFrame({
        'Calculated': pd_calcs['VIX'],
        'Actual': df_vix,
    })
    df_merged.plot(figsize=(10, 6), grid=True, style=['b', 'ro']);
```

This gives us the following output:

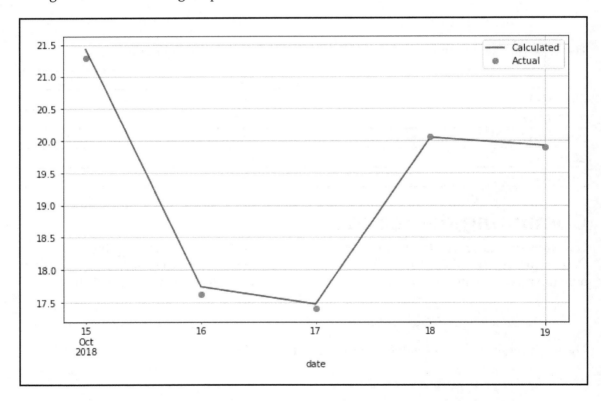

The calculated values in red dots appear to be very close to the actual VIX values.

Summary

In this chapter, we looked at volatility derivatives and their uses by investors to diversify and hedge their risk in equity and credit portfolios. Since long-term investors in equity funds are exposed to downside risk, volatility can be used as a hedge for the tail risk and in replacement for the put options. In the United States, the CBOE VIX measures the short-term volatility implied by SPX option prices. In Europe, the VSTOXX Market Index is based on the market prices of a basket of OESX, and measures the implied market volatility over the next 30 days on the EURO STOXX 50 Index. Many people around the world use the VIX as a popular measurement tool for the stock market volatility over the next 30-day period. To help us better understand how the VIX Index is calculated, we looked at its components and at formulas used in determining its value.

To help us determine the relationship between the SPX and the VIX, we downloaded these data and performed a variety of financial analytics, with the conclusion that they are negatively correlated. This relationship presents a viable way of avoiding frequent rebalancing costs by trading strategies based on benchmarking. The statistical nature of volatility allows volatility derivative traders to generate returns by utilizing mean-reverting strategies, dispersion trading, and volatility spread trading, among others.

In studying VIX-based trading strategies, we replicated the VIX Index for a single time period. Since the VIX Index is a sentiment of the volatility outlook for the next 30 days, it is made up of two SPX option chains expiring between 24 and 36 days. As the SPX rises and falls, the volatility of SPX options changes, and option bids may become zero. The number of options included in the calculation of the VIX Index may change as a result. For the simplicity of breaking down the VIX calculations in this chapter, we assumed the number of options included is static. We also assumed the CMTs over the 5-day period is constant. In reality, options prices and the risk-free interest rate are constantly changing, and the VIX Index recalculates approximately every 15 seconds.

In the next section, we will be building an algorithmic trading platform.

Building an Algorithmic Trading Platform

8

Algorithmic trading automates the systematic trading process, where orders are executed at the best price possible based on a variety of factors, such as pricing, timing, and volume. Brokerage firms may offer an **Application Programming Interface** (**API**) as part of their service offering for customers who wish to deploy their own trading algorithms. An algorithmic trading system must be highly robust to handle any point of failure during the order execution. Network configuration, hardware, memory management, speed, and user experience are a number of factors to be considered when designing a system for executing orders. Designing larger systems inevitably adds more complexity to the framework.

As soon as a position in a market is opened, it is subjected to various types of risk, such as market risk, interest rate risk, and liquidity risk. To preserve the trading capital as much as possible, it is important to incorporate risk management measures into the trading system. Perhaps the most common risk measure used in the financial industry is the **Value at Risk** (**VaR**) technique. We will discuss the beauty and flaws of VaR, and how it can be incorporated into our trading system that we will develop in this chapter.

In this chapter, we will cover the following topics:

- An overview of algorithmic trading
- A list of brokers and system vendors with a public API
- Choosing a programming language for a trading system
- Designing an algorithmic trading platform
- Setting up API access on the Oanda v20 Python module
- Implementing a mean-reverting algorithmic trading strategy
- Implementing a trend-following algorithmic trading strategy
- Introducing VaR for risk management in our trading system
- Performing VaR calculations in Python on AAPL

Introducing algorithmic trading

In the 1990s, exchanges had already begun to use electronic trading systems. By 1997, 44 exchanges worldwide used automated systems for trading futures and options with more exchanges in the process of developing automated technology. Exchanges such as the **Chicago Board of Trade** (**CBOT**) and the **London International Financial Futures and Options Exchange** (**LIFFE**) used their electronic trading systems as an after-hours complement to the traditional open outcry trading pits, thus giving traders 24-hour access to the exchange's risk management tools. With these improvements in technology, technology-based trading became less expensive, fueling the growth of trading platforms that are faster and more powerful. The higher reliability of order execution and the lower rate of message transmission error has deepened the reliance on technology by financial institutions. The majority of asset managers, proprietary traders, and market makers have since moved from the trading pits to electronic trading floors.

As systematic or computerized trading becomes more commonplace, speed is the most important factor in determining the outcome of a trade. Quants, by utilizing sophisticated fundamental models, are able to recompute fair values of trading products on the fly and execute trading decisions, enabling them to reap profits at the expense of those fundamental traders using traditional tools. This has given way to the term **high-frequency trading** (**HFT**), which relies on fast computers to execute the trading decisions before anyone else can. In fact, HFT has evolved into a billion-dollar industry.

Algorithmic trading refers to the automation of the systematic trading process, where the order execution is heavily optimized to give the best price possible. It is not part of the portfolio allocation process.

Banks, hedge funds, brokerage firms, clearing firms, and trading firms typically have their servers placed right next to the electronic exchange to receive the latest market prices and to perform the fastest order execution where possible. They bring enormous trading volumes to the exchange. Anyone who wishes to participate in low-latency, high-volume trading activities (such as complex event processing or capturing fleeting price discrepancies) by acquiring exchange connectivity may do so in the form of co-location, where their server hardware can be placed on a rack right next to the exchange for a fee.

The **Financial Information Exchange** (**FIX**) protocol is the industry standard for electronic communications with the exchange from the private server for **direct market access** (**DMA**) to real-time information. C++ is the common choice of programming language for trading over the FIX protocol, though other languages, such as .NET Framework Common Language and Java can also be used. The **Representational State Transfer(REST)** API offerings are becoming more common for retail investors. Before creating an algorithmic trading platform, you will need to assess various factors, such as the speed and ease of learning before deciding on a specific language for the purpose.

Brokerage firms will provide a trading platform of some sort for their customers in order for them to execute orders on selected exchanges in return for the commission fees. Some brokerage firms may offer an API as part of their service offering to technically-inclined customers who wish to run their own trading algorithms. In most circumstances, customers may also choose from a number of commercial trading platforms offered by third-party vendors. Some of these trading platforms may also offer API access to route orders electronically to the exchange. It is important to read the API documentation beforehand to understand the technical capabilities offered by your broker and to formulate an approach in developing an algorithmic trading system.

Trading platforms with a public API

The following table lists some brokers and trading platform vendors who have their API documentation publicly available:

Broker/vendor	URL	Programming languages supported
CQG	`https://www.cqg.com`	REST, FIX, C#, C++, and VB/VBA
Cunningham Trading Systems	`http://www.ctsfutures.com`	Microsoft .Net Framework 4.0 and FIX
E*Trade	`https://developer.etrade.com/home`	Python, Java, and Node.js
Interactive Brokers	`https://www.interactivebrokers.com/en/index.php?f=5041`	Java, C++, Python, C#, C++, and DDE
IG	`https://labs.ig.com/`	REST, Java, JavaScript, .NET, Clojure, and Node.js
Tradier	`https://developer.tradier.com/`	REST
Trading Technologies	`https://www.tradingtechnologies.com/trading/apis/`	REST, .NET, and FIX
OANDA	`https://developer.oanda.com/`	REST, Java, and FIX

Broker/vendor	URL	Programming languages supported
FXCM	`https://www.fxcm.com/uk/algorithmic-trading/api-trading/`	REST, Java, and FIX

Choosing a programming language

With many options of programming languages available to interface with brokers or vendors, the question that comes naturally to anyone starting out in algorithmic trading platform development is *Which language should I use?*

Before answering this question, it is important to find out if your broker provides developer tools. RESTful APIs are becoming the most common offering alongside FIX protocol access. A small number of brokers support Java and C#. With RESTful APIs, it is easy to search for, or even write a wrapper around it in almost any programming language that supports the **HyperText Transfer Protocol (HTTP)**.

Bear in mind that each tool option presents its own limitations. Your broker may rate-limit price and event updates. How your product will be developed, the performance metrics to follow, the costs involved, latency threshold, risk measures, and the expected user interface are pieces of the puzzle to be taken into consideration. The risk manager, execution engine, and portfolio optimizer are some major components that will affect the design of your system. Your existing trading infrastructure, choice of operating system, programming language compiler capability, and available software tools pose further constraints on the system design, development, and deployment.

System functionalities

It is important to define the outcomes of your trading system. An outcome could be a research-based system concerned with obtaining high-quality data from data vendors, performing computations or running models, and evaluating a strategy through signal generation. Part of the research component might include a data-cleaning module or a backtesting interface to run a strategy with theoretical parameters over historical data. The CPU speed, memory size, and bandwidth are factors to be considered while designing our system.

Another outcome could be an execution-based system that is more concerned with risk management and order handling features to ensure the timely execution of multiple orders. The system must be highly robust in order to handle any point of failure during the order execution. As such, network configuration, hardware, memory management and speed, and user experience are some factors to be considered when designing a system that executes orders.

A system may contain one or more of these functionalities. Designing larger systems inevitably adds complexity to the framework. It is recommended that you choose one or more programming languages that can address and balance the development speed, ease of development, scalability, and reliability of your trading system.

Building an algorithmic trading platform

In this section, we will design and build a live algorithmic trading system in Python. Since developer tools and offerings vary with each broker, it is important to take into consideration the different programming implementation that is required in integrating with our very own trading system. With a good system design, we can build a generic service that allows configurations of different brokers to plug in and play together nicely with our trading system.

Designing a broker interface

When designing a trading platform, the following three functionalities are highly desirable to fulfill any given trading plan:

- **Getting prices**: Pricing data is the one of the most basic units of information available from an exchange. It represents the quotation prices that are made by the marketplace to buy or sell a traded product. A broker may redistribute data from the exchange to be delivered to you in its own format. The most basic form of price data available is the date and time of the quotes, the symbol of the traded product, and the quoted bidding and asking price of the traded product. More often than not, this pricing data is useful to base your trading decisions on.

The best quoted bidding and asking prices are known as the **Level 1** quote. In most cases, it is possible to request for Level 2, 3, or even additional quote levels from your broker.

- **Sending orders to the market**: When sending orders to the marketplace, it may or may not be executed by your broker or the exchange. If it does get executed, you will have opened a position in the traded product and subjected yourself to all forms of risk, as well as returns. The simplest form of order states the product to trade (usually denoted by a symbol), the quantity to trade, the position that you want to take (that is, whether you are buying or selling), and, for a non-market order, the price to trade at. Depending on your needs, there are many different types of orders available to help manage your trading risks.

   Your broker may not support all order types. It is prudent to check with your broker which types of order are available and which can best manage your trading risks. The most common order type used by market participants are market orders, limit orders, and good-till-canceled orders. A **market order** is an order to buy or sell a product right away in the market. Because it is executed based on the current market prices, an execution price is not required for this type of order. A **limit order** is an order to buy or sell a product at a specific or better price. A **good-till-canceled** (**GTC**) order is an order that remains in the exchange queue for execution until the stated expiry time. Unless specified, most orders are GTC orders that expire at the end of the trading day. You can find more information of various order types at https://www.investopedia.com/university/how-start-trading/how-start-trading-order-types.asp.

- **Tracking positions:** As soon as your order is executed, you will enter into a position. Keeping track of your opened position will help determine how well (or badly!) your trading strategy is doing, as well as manage and plan your risks. Your gains and losses from opened positions vary according to market movements, and is known as **unrealized profits and losses**. After closing your position, you will have **realized profits and losses**, which are the final outcome of your trading strategy.

With these three basic functionalities in mind, we can design a generic `Broker` class implementing these functions that can be easily extended to any broker-specific configurations.

Python library requirements

In this chapter, we will be using the publicly-available v20 module with Oanda as our broker. All method implementations mentioned in this chapter uses the `v20` Python library as an example.

Installing v20

The official repository for OANDA v20 REST API is at `https://github.com/oanda/v20-python`. Install using pip with the terminal command:

```
pip install v20
```

Detailed documentation on the use of the OANDA v20 REST API can be found at `http://developer.oanda.com/rest-live-v20/introduction/`. The use of APIs varies with each broker, so make sure that you consult with your broker for the appropriate documentation before writing your trading system implementation.

Writing an event-driven broker class

Whether we are fetching prices, sending orders, or tracking positions, an event-driven system design will trigger key parts of our system in a multi-threaded fashion without blocking the main thread.

Let's begin writing our `Broker` class in Python, as follows:

```python
from abc import abstractmethod


class Broker(object):
    def __init__(self, host, port):
        self.host = host
        self.port = port

        self.__price_event_handler = None
        self.__order_event_handler = None
        self.__position_event_handler = None
```

In the constructor, we can provide the `host` and `port` public connection configurations of our broker for the inheriting child classes. Three variables are declared for storing the event handlers of prices, orders, and position updates, respectively. Here, we are designing for only one listener for each event. A more complex trading system might support multiple listeners on the same event handler.

Storing the price event handler

Inside the `Broker` class, add the following two methods as the getter and setter for the price event handler, respectively:

```
@property
def on_price_event(self):
    """
    Listeners will receive: symbol, bid, ask
    """
    return self.__price_event_handler

@on_price_event.setter
def on_price_event(self, event_handler):
    self.__price_event_handler = event_handler
```

The inheriting child classes will notify the listeners through the `on_price_event` method invocation of the symbol, the bid price, and the ask price. Later on, we will use these pieces of basic information for our trading decisions.

Storing the order event handler

Add the following two methods as the getter and setter for the order event handler, respectively:

```
@property
def on_order_event(self):
    """
    Listeners will receive: transaction_id
    """
    return self.__order_event_handler

@on_order_event.setter
def on_order_event(self, event_handler):
    self.__order_event_handler = event_handler
```

After an order is routed to your broker, the inheriting child classes will notify the listeners through the `on_order_event` method invocation, along with the order transaction ID.

Storing the position event handler

Add the following two methods as the getter and setter for the position event handler:

```
@property
def on_position_event(self):
    """
    Listeners will receive:
    symbol, is_long, units, unrealized_pnl, pnl
    """
    return self.__position_event_handler

@on_position_event.setter
def on_position_event(self, event_handler):
    self.__position_event_handler = event_handler
```

When a position update event is received from your broker, the inheriting child classes will notify the listeners through the `on_position_event` method invocation containing the symbol information, a flag indicating a long or short position, the number of units traded, the unrealized profit and loss, and the realized profit and loss.

Declaring an abstract method for getting prices

Since fetching prices from a data source is a main requirement of any trading system, create an abstract method named `get_prices()` to perform such a function. It expects a `symbols` parameter to contain a list of broker-defined symbols, that will be used for querying data from our broker. The inheriting child classes are expected to implement this method, otherwise a `NotImplementedError` exception is thrown:

```
@abstractmethod
def get_prices(self, symbols=[]):
    """
    Query market prices from a broker
    :param symbols: list of symbols recognized by your broker
    """
    raise NotImplementedError('Method is required!')
```

Note that this `get_prices()` method is expected to perform a one-time fetch of the current market prices. This gives us a snapshot of the market at a particular time. For a continuously-running trading system, we will require streaming market prices to feed our trading logic in real time, which we will define next.

Declaring an abstract method for streaming prices

Add a `stream_prices()` abstract method that accepts a list of symbols in streaming prices using the following code:

```
@abstractmethod
def stream_prices(self, symbols=[]):
    """
    Continuously stream prices from a broker.
    :param symbols: list of symbols recognized by your broker
    """
    raise NotImplementedError('Method is required!')
```

The inheriting child classes are expected to implement this method in streaming prices from your broker, otherwise a `NotImplementedError` exception message will be thrown.

Declaring an abstract method for sending orders

Add a `send_market_order()` abstract method for the inheriting child classes to implement when sending a market order to your broker:

```
@abstractmethod
def send_market_order(self, symbol, quantity, is_buy):
    raise NotImplementedError('Method is required!')
```

Using the preceding methods written in our `Broker` base class, we can now write broker-specific classes in the next section.

Implementing the broker class

In this section, we will implement the abstract methods of the `Broker` class that are specific to our broker, Oanda. This requires the use of the `v20` library. However, you can easily change the configuration and any implemented methods that are specific to a broker of your choice.

Initializing the broker class

Write the following `OandaBroker` class, which is specific to our broker, extending the generic `Broker` class:

```python
import v20

class OandaBroker(Broker):
    PRACTICE_API_HOST = 'api-fxpractice.oanda.com'
    PRACTICE_STREAM_HOST = 'stream-fxpractice.oanda.com'

    LIVE_API_HOST = 'api-fxtrade.oanda.com'
    LIVE_STREAM_HOST = 'stream-fxtrade.oanda.com'

    PORT = '443'

    def __init__(self, accountid, token, is_live=False):
        if is_live:
            host = self.LIVE_API_HOST
            stream_host = self.LIVE_STREAM_HOST
        else:
            host = self.PRACTICE_API_HOST
            stream_host = self.PRACTICE_STREAM_HOST

        super(OandaBroker, self).__init__(host, self.PORT)

        self.accountid = accountid
        self.token = token

        self.api = v20.Context(host, self.port, token=token)
        self.stream_api = v20.Context(stream_host, self.port, token=token)
```

Note that Oanda uses two different hosts for regular API endpoints and a streaming API endpoint. These endpoints are different for their practice and live trading environments. All endpoints are connected on the standard **Secure Socket Layer** (**SSL**) port 440. In the constructor, the `is_live` Boolean flag chooses the appropriate endpoints for the chosen trading environment for saving in the parent class. A `True` value for `is_live` indicates a live trading environment. The constructor argument also saves the account ID and token, which are required for authenticating the account used for trading. This information can be obtained from your broker.

The `api` and `stream_api` variables hold the `v20` library's `Context` objects that are used by calling methods to send instructions to your broker.

Implementing the method for getting prices

The following codes implement the parent `get_prices()` method in the `OandaBroker` class for getting prices from your broker:

```
def get_prices(self, symbols=[]):
    response = self.api.pricing.get(
        self.accountid,
        instruments=",".join(symbols),
        snapshot=True,
        includeUnitsAvailable=False
    )
    body = response.body
    prices = body.get('prices', [])
    for price in prices:
        self.process_price(price)
```

The body of the response contains a `prices` attribute and a list of objects. Each item in the list is processed by the `process_price()` method. Let's implement this method in the `OandaBroker` class as well:

```
def process_price(self, price):
    symbol = price.instrument

    if not symbol:
        print('Price symbol is empty!')
        return

    bids = price.bids or []
    price_bucket_bid = bids[0] if bids and len(bids) > 0 else None
    bid = price_bucket_bid.price if price_bucket_bid else 0

    asks = price.asks or []
    price_bucket_ask = asks[0] if asks and len(asks) > 0 else None
    ask = price_bucket_ask.price if price_bucket_ask else 0

    self.on_price_event(symbol, bid, ask)
```

The `price` object contains an `instrument` property of a string object, along with `list` objects in the `bids` and `asks` properties. Typically, Level 1 quotes are available, so we read the first item of the each list. Each item in the list is a `price_bucket` object, from which we extract the bid and ask price.

With this information extracted, we pass it to the `on_price_event()` event handler method. Note that, in this example, we are passing only three values. In more complex trading systems, you might want to consider extracting more detailed information, such as traded volume, last traded price, or multilevel quotes, and pass this to the price event listeners.

Implementing the method for streaming prices

Add the following `stream_prices()` method in the `OandaBroker` class to start streaming prices from your broker:

```
def stream_prices(self, symbols=[]):
    response = self.stream_api.pricing.stream(
        self.accountid,
        instruments=",".join(symbols),
        snapshot=True
    )

    for msg_type, msg in response.parts():
        if msg_type == "pricing.Heartbeat":
            continue
        elif msg_type == "pricing.ClientPrice":
            self.process_price(msg)
```

Since the host connection expects a continuous stream, the `response` object has a `parts()` method that listens for incoming data. The `msg` object is essentially a `price` object, which we can reuse with the `process_price()` method to notify the listeners of an incoming price event.

Implementing the method for sending market orders

Add the following `send_market_order()` method in the `OandaBroker` class, which will send a market order to your broker:

```
def send_market_order(self, symbol, quantity, is_buy):
    response = self.api.order.market(
        self.accountid,
        units=abs(quantity) * (1 if is_buy else -1),
        instrument=symbol,
        type='MARKET',
    )
    if response.status != 201:
        self.on_order_event(symbol, quantity, is_buy, None, 'NOT_FILLED')
        return

    body = response.body
    if 'orderCancelTransaction' in body:
```

```
        self.on_order_event(symbol, quantity, is_buy, None, 'NOT_FILLED')
        return

    transaction_id = body.get('lastTransactionID', None)
    self.on_order_event(symbol, quantity, is_buy, transaction_id, 'FILLED')
```

When the `market()` method of the v20 `order` library is called, the status of the response is expected to be `201` to indicate a successful connection to the broker. A further check on the response body is recommended for signs of error in the execution of our orders. In the case of a successful execution, the transaction ID and the details of the order are passed along to the listeners by calling the `on_order_event()` event handler. Otherwise, the order event is triggered with an empty transaction ID along with a `NOT_FILLED` status, indicating that the order is incomplete.

Implementing the method for fetching positions

Add the following `get_positions()` method in the `OandaBroker` class, which will fetch all the available position information for a given account:

```python
def get_positions(self):
    response = self.api.position.list(self.accountid)
    body = response.body
    positions = body.get('positions', [])
    for position in positions:
        symbol = position.instrument
        unrealized_pnl = position.unrealizedPL
        pnl = position.pl
        long = position.long
        short = position.short

        if short.units:
            self.on_position_event(
                symbol, False, short.units, unrealized_pnl, pnl)
        elif long.units:
            self.on_position_event(
                symbol, True, long.units, unrealized_pnl, pnl)
        else:
            self.on_position_event(
                symbol, None, 0, unrealized_pnl, pnl)
```

In the response body, the `position` property contains a list of `position` objects, each having attributes for the contract symbol, the unrealized and realized gains or losses, and the number of long and short positions. This information is passed along to the listeners through the `on_position_event()` event handler.

Getting the prices

With the methods of `broker` now defined, we can test the connection that is set up between our broker by reading the current market prices. The `Broker` class may be instantiated using the following Python codes:

```python
# Replace these 2 values with your own!
ACCOUNT_ID = '101-001-1374173-001'
API_TOKEN = '6ecf6b053262c590b78bb8199b85aa2f-
d99c54aecb2d5b4583a9f707636e8009'

broker = OandaBroker(ACCOUNT_ID, API_TOKEN)
```

Replace the two constant variables, `ACCOUNT_ID` and `API_TOKEN`, with your own credentials given by your broker, which identifies your own trading account. The `broker` variable is an instance of `OandaBroker`, which we can use to perform various broker-specific calls.

Suppose that we are interested in finding out the current market price of the EUR/USD currency pair. Let's define a constant variable to hold the symbol of this instrument that is recognized by our broker:

```python
SYMBOL = 'EUR_USD'
```

Next, define a price event listener coming from our broker, using the following code:

```python
import datetime as dt

def on_price_event(symbol, bid, ask):
    print(
        dt.datetime.now(), '[PRICE]',
        symbol, 'bid:', bid, 'ask:', ask
    )

broker.on_price_event = on_price_event
```

The `on_price_event()` function is defined as the listener for incoming price information and is assigned to the `broker.on_price_event` event handler. We expect three values from a pricing event – the contract symbol, the bid price, and the ask price – which we simply print to the console.

The `get_prices()` method is called to fetch the current market price from our broker:

```python
broker.get_prices(symbols=[SYMBOL])
```

We should get a similar output on the console as follows:

```
2018-11-19 21:29:13.214893 [PRICE] EUR_USD bid: 1.14361 ask: 1.14374
```

The output is a single line showing the bid and ask price of the EUR/USD currency pair as `1.14361` and `1.14374`, respectively.

Sending a simple market order

In the same way that we use for fetching prices, we can reuse the `broker` variable to send a market order to our broker.

Now suppose that we are interested in buying one unit of the same EUR/USD currency pair; the following code performs this action:

```python
def on_order_event(symbol, quantity, is_buy, transaction_id, status):
    print(
        dt.datetime.now(), '[ORDER]',
        'transaction_id:', transaction_id,
        'status:', status,
        'symbol:', symbol,
        'quantity:', quantity,
        'is_buy:', is_buy,
    )

broker.on_order_event = on_order_event
broker.send_market_order(SYMBOL, 1, True)
```

The `on_order_event()` function is defined as the listener for incoming order updates from our broker and is assigned to the `broker.on_order_event` event handler. For example, a limit order that is executed, or an order that is canceled, will be called on by this method. Finally, the `send_market_order()` method indicates that we are interested in buying one unit of the EUR/USD currency pair.

If the currency markets are open when you run the preceding codes, you should get the following result with a different transaction ID:

```
2018-11-19 21:29:13.484685 [ORDER] transaction_id: 754 status: FILLED
symbol: EUR_USD quantity: 1 is_buy: True
```

The output shows that the order is successfully filled to buy one unit of the EUR/USD currency pair with a transaction ID of `754`.

Getting position updates

With a long position opened by sending a market order to buy, we should be able to view our current EUR/USD position. We can do so on the `broker` object using the following code:

```
def on_position_event(symbol, is_long, units, upnl, pnl):
    print(
        dt.datetime.now(), '[POSITION]',
        'symbol:', symbol,
        'is_long:', is_long,
        'units:', units,
        'upnl:', upnl,
        'pnl:', pnl
    )

broker.on_position_event = on_position_event
broker.get_positions()
```

The `on_position_event()` function is defined as the listener for incoming position updates from our broker and is assigned to the `broke.on_position_event` event handler. When the `get_positions()` method is called, the broker returns the position information and triggers the following output:

```
2018-11-19 21:29:13.752886 [POSITION] symbol: EUR_USD is_long: True units:
1.0 upnl: -0.0001 pnl: 0.0
```

Our position statement is currently one long unit of the EUR/USD currency pair, with an unrealized loss of $0.0001. Since this is our first trade, we have not realized any profits or losses yet.

Building a mean-reverting algorithmic trading system

With our broker now accepting orders and responding to our requests, we can begin to design a fully-automated trading system. In this section, we will explore how to design and implement a mean-reverting algorithmic trading system.

Designing the mean-reversion algorithm

Suppose we believe that in normal market conditions, prices fluctuate, but tend to revert back to some short-term level, such as the average of the most recent prices. In this example, we assume that the EUR/USD currency pair is exhibiting a mean-reversion property in the near short-term period. First, we resample the raw tick-level data into standard time series intervals, for example, at one-minute intervals. Then, taking a number of the most recent periods for calculating the short-term average price (for example, with five periods), we are saying that we believe the EUR/USD prices will revert toward the average of the prior five minutes' prices.

As soon as the bidding price of the EUR/USD currency pair exceeds the short-term average price, with five minutes as our example, our trading system shall generate a sell signal, and we can choose to enter into a short position with a sell market order. Likewise, when the asking price of EUR/USD falls below the average price, a buy signal is generated and we can choose to enter into a long position with a buy market order.

The moment that a position is opened, we may use the same signals to close out our position. When a long position is opened, we close our position on a sell signal by entering an order to sell at the market. Likewise, when a short position is opened, we close our position on a buy signal by entering an order to buy at the market.

You might observe that there are plenty of flaws in our trading strategy. Closing out our position does not guarantee a profit. Our belief of the market can be wrong; in adverse market conditions, a signal might remain in one direction for some time and there is a high probability of closing out our position at a huge loss! As a trader, you should figure out a personal trading strategy that suits your beliefs and risk appetite.

Implementing the mean-reversion trader class

The resample interval and the number of periods in our calculation are two important parameters that are required by our trading system. First, create a class named `MeanReversionTrader`, which we can instantiate and run as our trading system:

```
import time
import datetime as dt
import pandas as pd


class MeanReversionTrader(object):
    def __init__(
        self, broker, symbol=None, units=1,
        resample_interval='60s', mean_periods=5
```

```
    ):
        """
        A trading platform that trades on one side
            based on a mean-reverting algorithm.

        :param broker: Broker object
        :param symbol: A str object recognized by the broker for trading
        :param units: Number of units to trade
        :param resample_interval:
            Frequency for resampling price time series
        :param mean_periods: Number of resampled intervals
            for calculating the average price
        """
        self.broker = self.setup_broker(broker)

        self.resample_interval = resample_interval
        self.mean_periods = mean_periods
        self.symbol = symbol
        self.units = units

        self.df_prices = pd.DataFrame(columns=[symbol])
        self.pnl, self.upnl = 0, 0

        self.mean = 0
        self.bid_price, self.ask_price = 0, 0
        self.position = 0
        self.is_order_pending = False
        self.is_next_signal_cycle = True
```

The five parameters in our constructor initialize the state of our trading system – the broker used, the symbol to trade, the number of units to trade, the resampling interval of our price data, and the number of periods for our mean calculation. These values are simply stored as class variables.

The setup_broker() method call sets up our class to handle events from our broker object, which we will define shortly. As we receive price data, these are stored in a pandas DataFrame variable, df_prices. The latest bid and ask prices are stored in the bid_price and ask_price variables for calculating signals. The mean variable will store the calculated mean of the prior number of mean_period prices. The position variable will store the number of units of our current position. A negative value indicates a short position, and a positive value indicates a long position.

The `is_order_pending` Boolean flag indicates whether an order is pending execution by our broker, and the `is_next_signal_cycle` Boolean flag indicates whether the current trading state cycle is open. Note that our system states can be as follows:

1. Wait for a buy or sell signal.
2. Place an order on a buy or sell signal.
3. When a position is opened, wait for a sell or buy signal.
4. Place an order on a sell or buy signal.
5. When the position is closed, go to step 1.

For every cycle of steps from 1 to 5, we will only trade one unit. These Boolean flags act as a lock to prevent multiple orders from entering into the system at any one time.

Adding event listeners

Let's hook up the price, order, and position events in our `MeanReversionTrader` class.

Add the `setup_broker()` method into this class, as follows:

```
def setup_broker(self, broker):
    broker.on_price_event = self.on_price_event
    broker.on_order_event = self.on_order_event
    broker.on_position_event = self.on_position_event
    return broker
```

We are simply assigning three class methods as listeners on any broker-generated event to listen to price, order, and position updates.

Add the `on_price_event()` method into this class, as follows:

```
def on_price_event(self, symbol, bid, ask):
    print(dt.datetime.now(), '[PRICE]', symbol, 'bid:', bid, 'ask:', ask)

    self.bid_price = bid
    self.ask_price = ask
    self.df_prices.loc[pd.Timestamp.now(), symbol] = (bid + ask) / 2.

    self.get_positions()
    self.generate_signals_and_think()

    self.print_state()
```

When a price event is received, we store them in our `bid_price`, `ask_price`, and `df_prices` class variables. As the price changes, so do our open positions and signal values. The `get_position()` method call retrieves up-to-date information on our positions, and the `generate_signals_and_think()` call recalculates our signals and decides whether to make the trade. The current state of the system is printed to the console using the `print_state()` command.

Write the `get_position()` method to retrieve the position information from our broker, as follows:

```
def get_positions(self):
    try:
        self.broker.get_positions()
    except Exception as ex:
        print('get_positions error:', ex)
```

Add the `on_order_event()` method into our class, as follows:

```
def on_order_event(self, symbol, quantity, is_buy, transaction_id, status):
    print(
        dt.datetime.now(), '[ORDER]',
        'transaction_id:', transaction_id,
        'status:', status,
        'symbol:', symbol,
        'quantity:', quantity,
        'is_buy:', is_buy,
    )
    if status == 'FILLED':
        self.is_order_pending = False
        self.is_next_signal_cycle = False

        self.get_positions()  # Update positions before thinking
        self.generate_signals_and_think()
```

When an order event is received, we print them out to the console. In our broker's `on_order_event` implementation, an order that is executed successfully will pass either a `status` value of `FILLED` or `UNFILLED`. Only on a successful order can we turn off our Boolean locks, retrieve our latest position, and perform decision-making for closing out our position.

Add the on_position_event() method into our class, as follows:

```
def on_position_event(self, symbol, is_long, units, upnl, pnl):
    if symbol == self.symbol:
        self.position = abs(units) * (1 if is_long else -1)
        self.pnl = pnl
        self.upnl = upnl
        self.print_state()
```

When a position update event is received for our intended trade symbol, we store our position information, the realized gains, and unrealized gains. The current state of the system is printed to the console using the print_state() command.

Add the print_state() method into our class, as follows:

```
def print_state(self):
    print(
        dt.datetime.now(), self.symbol, self.position_state,
        abs(self.position), 'upnl:', self.upnl, 'pnl:', self.pnl
    )
```

As soon as there are any updates to our orders, positions, or market prices, we print the latest state of our system to the console.

Writing the mean-reversion signal generators

We want our decision-making algorithm to recalculate trading signals on every price or order update. Let's create a generate_signals_and_think() method inside the MeanReversionTrader class to do this:

```
def generate_signals_and_think(self):
    df_resampled = self.df_prices\
        .resample(self.resample_interval)\
        .ffill()\
        .dropna()
    resampled_len = len(df_resampled.index)

    if resampled_len < self.mean_periods:
        print(
            'Insufficient data size to calculate logic. Need',
            self.mean_periods - resampled_len, 'more.'
        )
        return

    mean = df_resampled.tail(self.mean_periods).mean()[self.symbol]
```

```
# Signal flag calculation
is_signal_buy = mean > self.ask_price
is_signal_sell = mean < self.bid_price

print(
    'is_signal_buy:', is_signal_buy,
    'is_signal_sell:', is_signal_sell,
    'average_price: %.5f' % mean,
    'bid:', self.bid_price,
    'ask:', self.ask_price
)

self.think(is_signal_buy, is_signal_sell)
```

Since price data are stored in the df_prices variable as a pandas DataFrame, we can resample them at regular intervals, as defined by the resample_interval variable given in the constructor. The ffill() method forward-fills any missing data and the dropna() command removes the first missing value after resampling. There must be sufficient data available for calculating the mean, otherwise this method simply exits. The mean_periods variable represents the minimum length of resampled data that must be available.

The tail(self.mean_periods) method takes the most recent resampled intervals and calculates the average using the mean() method, resulting in another pandas DataFrame. The mean level is taken by index referencing the column of the DataFrame, which is simply the instrument symbol.

Using the average price available for the mean-reversion algorithm, we can generate the buy and sell signals. Here, a buy signal is generated when the average price exceeds the market asking price, and a sell signal is generated when the average price exceeds the market bidding price. Our short-term belief is that market prices will revert to the average price.

After printing these calculated values to the console for better debugging, we can now make use of the buy and sell signals to perform actual trades in a separate method named think() inside the same class:

```
def think(self, is_signal_buy, is_signal_sell):
    if self.is_order_pending:
        return

    if self.position == 0:
        self.think_when_flat_position(is_signal_buy, is_signal_sell)
    elif self.position > 0:
        self.think_when_position_long(is_signal_sell)
```

```
        elif self.position < 0:
            self.think_when_position_short(is_signal_buy)
```

If an order is still in pending state by a broker, we simply do nothing and exit the method. Since market conditions may change at any time, you might want to add your own logic to handle orders that have stayed in the pending state for too long and try another strategy.

The three if-else statements handles the trading logic when our position is flat, long, or short, respectively. When our position is flat, the `think_when_position_flat()` method is called, written as the following:

```
def think_when_position_flat(self, is_signal_buy, is_signal_sell):
    if is_signal_buy and self.is_next_signal_cycle:
        print('Opening position, BUY',
            self.symbol, self.units, 'units')
        self.is_order_pending = True
        self.send_market_order(self.symbol, self.units, True)
        return

    if is_signal_sell and self.is_next_signal_cycle:
        print('Opening position, SELL',
            self.symbol, self.units, 'units')
        self.is_order_pending = True
        self.send_market_order(self.symbol, self.units, False)
        return

    if not is_signal_buy and not is_signal_sell:
        self.is_next_signal_cycle = True
```

The first `if` statement handles the condition that, upon a buy signal and when the current trading cycle is open, we enter into a long position by sending a market order to buy and mark that order as pending. Conversely, the second `if` statement handles the condition to enter into a short position upon a sell signal. Otherwise, since the position is flat with neither a buy nor sell signal, we simply set the `is_next_signal_cycle` to `True` until a signal becomes available.

When we are in a long position, the `think_when_position_long()` method is called, written as the following:

```
def think_when_position_long(self, is_signal_sell):
    if is_signal_sell:
        print('Closing position, SELL',
            self.symbol, self.units, 'units')
        self.is_order_pending = True
        self.send_market_order(self.symbol, self.units, False)
```

On a sell signal, we mark the order as pending and close out our long position immediately by sending a market order to sell.

Similarly, when we are in a short position, the `think_when_position_short()` method is called, written as the following:

```python
def think_when_position_short(self, is_signal_buy):
    if is_signal_buy:
        print('Closing position, BUY',
                self.symbol, self.units, 'units')
        self.is_order_pending = True
        self.send_market_order(self.symbol, self.units, True)
```

On a buy signal, we mark the order as pending and close out our short position immediately by sending a market order to buy.

To perform the order routing functionality, add the following `send_market_order()` class method to our `MeanReversionTrader` class:

```python
def send_market_order(self, symbol, quantity, is_buy):
    self.broker.send_market_order(symbol, quantity, is_buy)
```

The order information is simply forwarded to our `Broker` class for execution.

Running our trading system

Finally, to start running our trading system, we need an entry point. Add the following `run()` class method to the `MeanReversionTrader` class:

```python
def run(self):
    self.get_positions()
    self.broker.stream_prices(symbols=[self.symbol])
```

During the first run of our trading system, we read our current positions and use the information to initialize all position-related information. Then, we request our broker to start streaming prices for the given symbol and hold the connection until the program is terminated.

With an entry point defined, all we need to do is initialize our `MeanReversionTrader` class and call the `run()` command using the following codes:

```python
trader = MeanReversionTrader(
    broker,
    symbol='EUR_USD',
    units=1
```

```
        resample_interval='60s',
        mean_periods=5,
)
trader.run()
```

Remember that the `broker` variable contains an instance of the `OandaBroker` class as defined from the previous *Getting prices* section, and we can reuse it for this class. Our trading system will use this broker object to perform broker-related calls. We are interested in the EUR/USD currency pair, trading one unit at each time. The `resample_interval` variable with a value of `60s` states that our stored prices are to be resampled at one-minute intervals. The `mean_periods` variable with a value of `5` states that we will take the average of the most recent five intervals, or the average price of the past five minutes.

To start our trading system, make the call to `run()`; pricing updates will start trickling in, enabling our system to trade on its own. You should see an output on the console that is similar to the following:

```
...
2018-11-21 15:19:34.487216 [PRICE] EUR_USD bid: 1.1393 ask: 1.13943
2018-11-21 15:19:35.686323 EUR_USD FLAT 0 upnl: 0.0 pnl: 0.0
Insufficient data size to calculate logic. Need 5 more.
2018-11-21 15:19:35.694619 EUR_USD FLAT 0 upnl: 0.0 pnl: 0.0
...
```

From the output, it looks as though our position is currently flat, and there is insufficient pricing data for calculating our trading signals.

After five minutes, when there is sufficient data for a trading signal calculation, we should be able to observe the following outcome:

```
...
2018-11-21 15:25:07.075883 EUR_USD FLAT 0 upnl: 0.0 pnl: -0.3246
is_signal_buy: False is_signal_sell: True average_price: 1.13934 bid:
1.13936 ask: 1.13949
Opening position, SELL EUR_USD 1 units
2018-11-21 15:25:07.356520 [ORDER] transaction_id: 2848 status: FILLED
symbol: EUR_USD quantity: 1 is_buy: False
2018-11-21 15:25:07.688082 EUR_USD SHORT 1.0 upnl: -0.0001 pnl: 0.0
is_signal_buy: False is_signal_sell: True average_price: 1.13934 bid:
1.13936 ask: 1.13949
2018-11-21 15:25:07.692292 EUR_USD SHORT 1.0 upnl: -0.0001 pnl: 0.0

...
```

The average price of the past five minutes is 1.13934. Since the current market bidding price for EUR/USD is 1.13936, more than the average price, a sell signal is generated. A sell market order is generated to open a short position in EUR/USD of one unit. This leads to an unrealized loss of $0.0001.

Let the system run on its own for a while, and it should be able to close out the positions on its own. To stop trading, terminate the running process using *Ctrl + Z* or something similar. Remember to manually close out any remaining trading positions once the program stops running. You now have a fully functional and automated trading system!

 The system design and trading parameters here are stated as an example and don't necessarily lead to positive outcomes! You should experiment with various trading parameters and improve the handling of events to figure out the optimal strategy for your trading plan.

Building a trend-following trading platform

In the previous section, we followed the steps for building a mean-reverting trading platform. The same functionality can be easily extended to incorporate any other trading strategies. In this section, we will take a look at reusing the MeanReversionTrader class to implement a trend-following trading system.

Designing the trend-following algorithm

Suppose that this time, we believe that the current market conditions exhibit a trend-following pattern, perhaps due to seasonal changes, economic projections, or government policy. As prices fluctuate, and as the short-term average price level crosses the average long-term price level by a certain threshold, we generate a buy or sell signal.

First, we resample raw tick-level data into standard time series intervals, for example, at one-minute intervals. Second, taking a number of the most recent periods, for example, with five periods, we calculate the short-term average price for the past five minutes. Finally, taking a larger number of the most recent periods, for example, with ten periods, we calculate the long-term average price for the past ten minutes.

In a market with no movement, the average short-term price should be the same as the average long-term price with a ratio of one – this ratio is also known as the beta. When the average short-term price increases more than the average long-term price, the beta is more than one and the market can be viewed as on an uptrend. When the short-term price decreases more than the average long-term price, the beta is less than one and the market can be viewed as on a downtrend.

On an uptrend, as soon as the beta crosses above a certain price threshold level, our trading system shall generate a buy signal, and we can choose to enter into a long position with a buy market order. Likewise, on a downtrend, when the beta falls below a certain price threshold level, a sell signal is generated and we can choose to enter into a short position with a sell market order.

The moment that a position is opened, the same signals may be used to close out our position. When a long position is opened, we close our position on a sell signal by entering an order to sell at the market. Likewise, when a short position is opened, we close out our position on a buy signal by entering an order to buy at the market.

The mechanics mentioned are very similar to those of the mean-reversion trading system design. Bear in mind that this algorithm does not guarantee any profits, and is just a simplistic belief of the markets. You should have a different (and better) view than this.

Writing the trend-following trader class

Let's write a class for our trend-following trading system with a new class named TrendFollowingTreader, which simply extends the MeanReversionTrader class using the following Python code:

```python
class TrendFollowingTrader(MeanReversionTrader):
    def __init__(
        self, *args, long_mean_periods=10,
        buy_threshold=1.0, sell_threshold=1.0, **kwargs
    ):
        super(TrendFollowingTrader, self).__init__(*args, **kwargs)

        self.long_mean_periods = long_mean_periods
        self.buy_threshold = buy_threshold
        self.sell_threshold = sell_threshold
```

In our constructor, we define three additional keyword arguments, `long_mean_periods`, `buy_threshold`, and `sell_threshold`, saved as class variables.

The `long_mean_periods` variable defines the number of resample intervals of our time series prices to take into account for calculating the long-term average price. Note that the existing `mean_periods` variable in the parent constructor is used for calculating the short-term average price. The `buy_threshold` and `sell_threshold` variables contain values that determine the boundaries of beta in generating a buy or sell signal.

Writing the trend-following signal generators

Because only the decision-making logic needs to be modified from our parent `MeanReversionTrader` class, and everything else, including orders, placement, and streaming prices, remains the same, we simply override the `generate_signals_and_think()` method and implement our new trend-following signal generators using the following code:

```
def generate_signals_and_think(self):
    df_resampled = self.df_prices\
        .resample(self.resample_interval)\
        .ffill().dropna()
    resampled_len = len(df_resampled.index)

    if resampled_len < self.long_mean_periods:
        print(
            'Insufficient data size to calculate logic. Need',
            self.mean_periods - resampled_len, 'more.'
        )
        return

    mean_short = df_resampled\
        .tail(self.mean_periods).mean()[self.symbol]
    mean_long = df_resampled\
        .tail(self.long_mean_periods).mean()[self.symbol]
    beta = mean_short / mean_long

    # Signal flag calculation
    is_signal_buy = beta > self.buy_threshold
    is_signal_sell = beta < self.sell_threshold

    print(
        'is_signal_buy:', is_signal_buy,
        'is_signal_sell:', is_signal_sell,
        'beta:', beta,
        'bid:', self.bid_price,
```

```
        'ask:', self.ask_price
    )

    self.think(is_signal_buy, is_signal_sell)
```

As before, on every invocation of the `generate_signals_and_think()` method, we resample prices at fixed intervals, defined by `resample_interval`. The minimum intervals required for the calculation of signals is now defined by `long_mean_periods` instead of `mean_periods`. The `mean_short` variable refers to the short-term average resampled price, and the `mean_long` variable refers to the long-term average resampled price.

The `beta` variable is the ratio of the short-term average price to the long-term average price. When the beta rises above the `buy_threshold` value, a buy signal is generated and the `is_signal_buy` variable is `True`. Likewise, when the beta falls below the `sell_threshold` value, a sell signal is generated and the `is_signal_sell` variable is `True`.

The trading parameters are printed to the console for debugging purposes, and the call to the parent `think()` class method triggers the usual logic of buying and selling with market orders.

Running the trend-following trading system

Let's start our trend-following trading system by instantiating the `TrendFollowingTrader` class and running it using the following code:

```
trader = TrendFollowingTrader(
    broker,
    resample_interval='60s',
    symbol='EUR_USD',
    units=1,
    mean_periods=5,
    long_mean_periods=10,
    buy_threshold=1.000010,
    sell_threshold=0.99990,
)
trader.run()
```

The first parameter, `broker`, is the same object created for our broker in the previous section. Again, we are resampling our time series prices at one-minute intervals, and we are interested in trading the EUR/USD currency pair, entering into a position of at most one unit at any given time. With a `mean_periods` value of 5, we are interested in taking the most recent five resampled intervals in calculating the average price of the past five minutes as our short-term average price. With a `long_mean_period` of 10, we are interested in taking the most recent 10 resampled intervals in calculating the average price of the past 10 minutes as our long-term average price.

The ratio of the short-term average price to the long-term average price is taken as the beta. When the beta rises above the value defined by `buy_threshold`, a buy signal is generated. When the beta falls below the value defined by `sell_threshold`, a sell signal is generated.

With our trading parameters set up, the `run()` method is called to start the trading system. We should see an output on the console that is similar to the following:

```
...
2018-11-23 08:51:12.438684 [PRICE] EUR_USD bid: 1.14018 ask: 1.14033
2018-11-23 08:51:13.520880 EUR_USD FLAT 0 upnl: 0.0 pnl: 0.0
Insufficient data size to calculate logic. Need 10 more.
2018-11-23 08:51:13.529919 EUR_USD FLAT 0 upnl: 0.0 pnl: 0.0
...
```

At the start of trading, we obtained the current market prices, staying in a flat position with neither profits nor losses. There is insufficient data available to make any trading decisions, and we will have to wait 10 minutes before we can see the calculated parameters take effect.

 If your trading system depends on a longer period of past data and you do not wish to wait for all this data to be collected, consider bootstrapping your trading system with historical data.

After a while, you should see an output that is similar to the following:

```
...
is_signal_buy: True is_signal_sell: False beta: 1.0000333228980047 bid:
1.14041 ask: 1.14058
Opening position, BUY EUR_USD 1 units
2018-11-23 09:01:01.579208 [ORDER] transaction_id: 2905 status: FILLED
symbol: EUR_USD quantity: 1 is_buy: True
2018-11-23 09:01:01.844743 EUR_USD LONG 1.0 upnl: -0.0002 pnl: 0.0
...
```

Let the system run on its own for awhile, and it should be able to close out positions on its own. To stop trading, terminate the running process with *Ctrl + Z*, or something similar. Remember to manually close out any remaining trading positions once the program stops running. Take steps to change your trading parameters and decision logic to make your trading system a profitable one!

 Note that the author is not responsible for any outcomes of your trading system! In a live trading environment, it takes more control parameters, order management, and position tracking to manage your risk effectively.

In the next section, we will discuss a risk management strategy that we can apply to our trading plans.

VaR for risk management

As soon as we open a position in the market, we are exposed to various types of risks, such as volatility risk and credit risk. To preserve our trading capital as much as possible, it is important to incorporate some form of risk management measures to our trading system.

Perhaps the most common measure of risk used in the financial industry is the VaR technique. It is designed to simply answer the following question: *What is the worst expected amount of loss, given a specific probability level, say 95%, over a certain period of time?* The beauty of VaR is that it can be applied to multiple levels, from position-specific micro-level to portfolio-based macro-level. For example, a VaR of $1 million with a 95% confidence level for a 1-day time horizon states that, on average, only 1 day out of 20 could you expect to lose more than $1 million due to market movements.

The following diagram illustrates a normally distributed portfolio returns with a mean of 0%, where VaR is the loss corresponding to the 95th percentile of the distribution of portfolio returns:

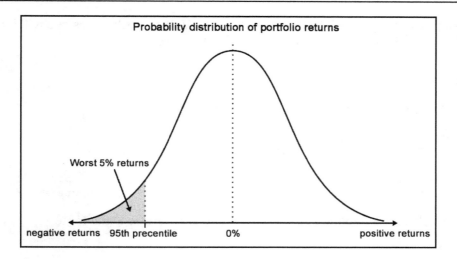

Suppose that we have $100 million under management at a fund claiming to have the same risk as an S&P 500 index fund, with an expected return of 9% and a standard deviation of 20%. To calculate the daily VaR at the 5% risk level or 95% confidence level using the variance-covariance method, we will use the following formulas:

$$daily\ volatility, \sigma = \frac{20\%}{\sqrt{252}} = 1.26\%$$

$$daily\ expected\ return, u = \frac{9\%}{252} = 0.036\%$$

$$VaR = P - P(N^{-1}(\alpha, u, \sigma) + 1) = \$2,036,606.50$$

Here, P is the value of the portfolio, and $N^{-1}(\alpha, u, \sigma)$ is the inverse normal probability distribution with a risk level of α, a mean of u, a and standard deviation of σ. The number of trading days per year is assumed to be 252. It turns out that the daily VaR at the 5% level is $2,036,606.50.

However, the use of VaR is not without its flaws. It does not take into account the probability of the loss for extreme events happening on the far ends of the tails on the normal distribution curve. The magnitude of the loss beyond a certain VaR level is difficult to estimate as well. The VaR that we investigated uses historical data and an assumed constant volatility level – such measures are not indicative of our future performance.

Let's take a practical approach to calculate the daily VaR of stock prices; we will investigate the AAPL stock prices by downloading from a data source:

```
"""
Download the all-time AAPL dataset
"""
from alpha_vantage.timeseries import TimeSeries

# Update your Alpha Vantage API key here...
ALPHA_VANTAGE_API_KEY = 'PZ2ISG9CYY379KLI'

ts = TimeSeries(key=ALPHA_VANTAGE_API_KEY, output_format='pandas')
df, meta_data = ts.get_daily_adjusted(symbol='AAPL', outputsize='full')
```

The dataset will be downloaded to the `df` variable as a pandas DataFrame:

```
df.info()
```

This gives us the following output:

```
<class 'pandas.core.frame.DataFrame'>
Index: 5259 entries, 1998-01-02 to 2018-11-23
Data columns (total 8 columns):
1. open                5259 non-null float64
2. high                5259 non-null float64
3. low                 5259 non-null float64
4. close               5259 non-null float64
5. adjusted close      5259 non-null float64
6. volume              5259 non-null float64
7. dividend amount     5259 non-null float64
8. split coefficient   5259 non-null float64
dtypes: float64(8)
memory usage: 349.2+ KB
```

Our DataFrame contains eight columns, with prices starting from the year 1998 to the present trading day. The column of interest is the adjusted closing prices. Suppose that we are interested in calculating the daily VaR of 2017; let's obtain this dataset using the following code:

```
import datetime as dt
import pandas as pd

# Define the date range
start = dt.datetime(2017, 1, 1)
end = dt.datetime(2017, 12, 31)

# Cast indexes as DateTimeIndex objects
df.index = pd.to_datetime(df.index)
```

```
closing_prices = df['5. adjusted close']
prices = closing_prices.loc[start:end]
```

The `prices` variable contains our AAPL dataset for 2017.

Using the formulas discussed earlier, you can implement the `calculate_daily_var()` function using the following code:

```
from scipy.stats import norm

def calculate_daily_var(
    portfolio, prob, mean,
    stdev, days_per_year=252.
):
    alpha = 1-prob
    u = mean/days_per_year
    sigma = stdev/np.sqrt(days_per_year)
    norminv = norm.ppf(alpha, u, sigma)
    return portfolio - portfolio*(norminv+1)
```

Let's assume that we are holding $100 million of AAPL stock, and we are interested in finding the daily VaR at the 95% confidence level. We can define the VaR parameters using the following code:

```
import numpy as np

portfolio = 100000000.00
confidence = 0.95

daily_returns = prices.pct_change().dropna()
mu = np.mean(daily_returns)
sigma = np.std(daily_returns)
```

The `mu` and `sigma` variables represent the daily mean percentage returns and the daily standard deviation of returns respectively.

We can obtain the VaR by calling the `calculate_daily_var()` function, as follows:

```
VaR = calculate_daily_var(
    portfolio, confidence, mu, sigma, days_per_year=252.)
print('Value-at-Risk: %.2f' % VaR)
```

We will get the following output:

```
Value-at-Risk: 114248.72
```

Assuming 252 trading days per year, the daily VaR of 2017 for the stock AAPL with 95% confidence is $114,248.72.

Summary

In this chapter, we were introduced to the evolution of trading from the pits to the electronic trading platform, and learned how algorithmic trading came about. We looked at some brokers offering API access to their trading service offering. To help us get started on our journey of developing an algorithmic trading system, we used the Oanda `v20` library to implement a mean-reversion trading system.

In designing an event-driven broker interface class, we defined event handlers for listening to orders, prices, and position updates. Child classes inheriting the `Broker` class simply extend this interface class with broker-specific functions, while still keeping the underlying trading functions compatible with our trading system. We successfully tested the connection with our broker by getting market prices, sending a market order, and receiving position updates.

We discussed the design of a simple mean-reversion trading system that generates buy or sell signals based the movements of historical average prices, and opening and closing out our positions with market orders. Since this trading system uses only one source of trading logic, more work will be required to build a robust, reliable, and profitable trading system.

We also discussed the design of a trend-following trading system that generates buy or sell signals based on the movements of a short-term average price against a long-term average price. With a well-designed system, we saw how easy it was to make a modification to the existing trading logic by simply extending the mean-reversion parent class and overriding the decision-making method.

One critical aspect of trading is to manage risk effectively. In the financial industry, VaR is the most common technique used to measure risk. Using Python, we took a practical approach to calculate the daily VaR of past datasets on AAPL.

Once we have built a working algorithmic trading system, we can explore the other ways to measure the performance of our trading strategy. One of these areas is backtesting; we will discuss this topic in the next chapter.

Implementing a Backtesting System 9

A **backtest** is a simulation of a model-driven investment strategy's response to historical data. While working on designing and developing a backtest, it would be helpful to think in terms of the concept of creating video games.

In this chapter, we will design and implement an event-driven backtesting system using an object-oriented approach. The resulting profits and losses of our trading model may be plotted on to a graph to help visualize the performance of our trading strategy. However, is this sufficient enough to determine whether it is a good model?

There are many concerns to be addressed in backtesting—for example, the effects of transaction costs, execution latency of orders, access to detailed transactions, and quality of historical data. Notwithstanding these factors, the primary goal of creating a backtesting system is to test a model as accurately as possible.

Backtesting involves a lot of research that merits its own literature. We will briefly cover some thoughts that you might want to consider when implementing a backtest. Typically, a number of algorithms are employed in backtesting. We will briefly discuss some of these: k-means clustering, k-nearest neighbors, classification and regression trees, 2k factorial design, and genetic algorithms.

In this chapter, we will cover the following topics:

- Introducing backtesting
- Concerns in backtesting
- Concept of an event-driven backtesting system
- Designing and implementing a backtesting system
- Writing classes to store tick data and market data
- Writing classes for orders and positions

- Writing a mean-reverting strategy
- Running the backtest engine single and multiple times
- Ten considerations for a backtesting model
- Discussion of algorithms in backtesting

Introducing backtesting

A backtest is a simulation of a model-driven investment strategy's response to historical data. The purpose of performing experiments with backtests is to make discoveries about a process or system. By using historical data, you can save time in testing an investment strategy for the period forward. It helps you test an investment theory based on the movements of the tested period. It is also used to both evaluate and calibrate an investment model. Creating a model is only the first step. The investment strategy will typically employ the model to help you drive simulated trading decisions and compute various factors related to either risk or return. These factors are typically used together to find a combination that is predictive of return.

Concerns in backtesting

However, there are many concerns to be addressed in backtesting:

- A backtest can never exactly replicate the performance of an investment strategy in an actual trading environment.
- The quality of the historical data is questionable, since it is subjected to outliers by third-party data vendors.
- Look-ahead bias takes many forms. For example, listed companies may split, merge, or de-list, resulting in substantial changes to its stock price.
- For strategies based on information from the order book, the market microstructure is extremely difficult to simulate realistically, since it represents the collective visible supply and demand in continuous time. This supply and demand are in turn affected by news events around the world.
- Icebergs and resting orders are some hidden elements of the market that could affect the structure once activated
- Other factors to be considered are transaction costs, execution latency of orders, and access to detailed transactions from backtesting.

Notwithstanding these factors, the primary goal of creating a backtesting system is thus to test a model as accurately as possible.

 Look-ahead bias is the use of available future data during the period it is being analyzed, resulting in inaccurate results in the simulation or study. It is vital to use information that would be only available during the period of study.

 In finance, iceberg orders are large orders that are broken up into several small orders. Only a small portion of the order is visible to the public—just like the *tip of the iceberg*—while the mass of the actual order is hidden. A **resting order** is an order whose price is away from the market and is waiting to be executed.

Concept of an event-driven backtesting system

While working on designing and developing a backtest, it would be helpful to think in terms of the concept of creating video games. After all, we are trying to create a simulated market pricing and ordering environment, very much akin to creating a virtual gaming world. Trading can also be regarded as a thrilling game of buying low and selling high.

In a virtual trading environment, components are needed for the simulation of price feeds, the order-matching engine, the order-book management, as well as functions for account and position updates. To achieve these functionalities, we can explore the concept of an event-driven backtesting system.

Let's start by understanding the concept of an event-driven programming paradigm used throughout the game development process. A system typically receives events as its inputs. It might be a keystroke entered by a user or a mouse movement. Other events could be messages that are generated by another system, a process, or a sensor to notify the host system of an incoming event.

The following diagram illustrates the stages involved in a game engine system:

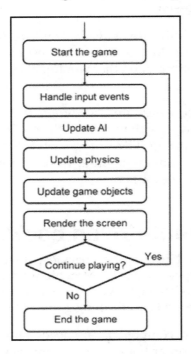

Let's take a look at a pseudo-code implementation of a main game engine loop:

```
while is_main_loop:  # Main game engine loop
    handle_input_events()
    update_AI()
    update_physics()
    update_game_objects()
    render_screen()
    sleep(1/60)  # Assuming a 60 frames-per-second video game rate
```

The core functions within the main game engine loop may process generated system events, as in the case of the `handle_input_events()` function, which handles the keyboard events:

```
def handle_input_events()
    event = get_latest_event()
    if event.type == 'UP_KEY_PRESS':
        move_player_up()
    elif event.type == 'DOWN_KEY_PRESS':
        move_player_down()
```

Using an event-driven system, such as the preceding example, helps us achieve code modularity and reusability by being able to swap and use similar events from different system components. The use of object-oriented programming is further enforced, where classes define objects in a game. These features are particularly useful for interfacing with different market data sources, multiple trading algorithms, and runtime environments when designing our trading platform. The simulated trading environment is close to being a realistic one and helps us prevent look-ahead bias.

Designing and implementing a backtesting system

Now that we have an idea of designing a video game for creating a backtesting trading system, we can begin our object-oriented approach by first defining the required classes for the various components in our trading system.

We are interested in implementing a simple backtesting system to test a mean-reverting strategy. Using the daily historical prices from a data-source provider, we will take the closing price of each day to compute the volatility of price returns for a particular instrument, using the AAPL stock price as an example. We would like to test a theory that if the standard deviation of returns for an elapsed number of days is far from the mean of zero by a particular threshold, a buy or sell signal is generated. When such a signal is indeed generated, a market order is sent to the exchange to be executed at the opening price of the next trading day.

As soon as we open a position, we would like to track our unrealized and realized profits to date. Our open position can be closed when an opposing signal is generated. On completion of the backtest, we will plot our profits and losses to see how well our strategy holds.

Does our theory sound like a viable trading strategy? Let's find out! The following sections explain the classes that go into implementing a backtesting system.

Writing a class to store tick data

Write a class named `TickData` that represents a single unit of data received from a market data source with the following Python code:

```python
class TickData(object):
    """ Stores a single unit of data """
```

```
        def __init__(self, timestamp='', symbol='',
                     open_price=0, close_price=0, total_volume=0):
            self.symbol = symbol
            self.timestamp = timestamp
            self.open_price = open_price
            self.close_price = close_price
            self.total_volume = total_volume
```

In this example, we are interested in storing the timestamp, symbol of the instrument, the opening and closing price, as well as the total volume traded. Detailed descriptions of a single unit of tick data, such as the highest price reached or last-traded volume, can be added as our system evolves.

Writing a class to store market data

An instance of the MarketData class is used throughout the system to store and retrieve prices referenced by various components. Essentially, it is a container for storing the last available tick data. Additional get helper functions are included to provide easy reference to the required information:

```
class MarketData(object):
    """ Stores the most recent tick data for all symbols """

    def __init__(self):
        self.recent_ticks = dict()   # indexed by symbol

    def add_tick_data(self, tick_data):
        self.recent_ticks[tick_data.symbol] = tick_data

    def get_open_price(self, symbol):
        return self.get_tick_data(symbol).open_price

    def get_close_price(self, symbol):
        return self.get_tick_data(symbol).close_price

    def get_tick_data(self, symbol):
        return self.recent_ticks.get(symbol, TickData())

    def get_timestamp(self, symbol):
        return self.recent_ticks[symbol].timestamp
```

Writing a class to generate sources of market data

Write a class called `MarketDataSource` to help us fetch historical data from an external data provider. In this example, we will be using **Quandl** as our data provider. The constructor of the class is defined as follows:

```
class MarketDataSource(object):
    def __init__(self, symbol, tick_event_handler=None, start='', end=''):
        self.market_data = MarketData()

        self.symbol = symbol
        self.tick_event_handler = tick_event_handler
        self.start, self.end = start, end
        self.df = None
```

In the constructor, the `symbol` parameter contains the value recognized by our data provider to download our desired dataset. An object of `MarketData` is instantiated for storing the most recent market data available. The `tick_event_handler` parameter stores the method handler as we iterate through our data source. The `start` and `end` parameters refer to the starting and ending dates of the dataset that we wish to keep in the pandas DataFrame variable, `df`.

Add the `fetch_historical_prices()` method inside the `MarketDataSource` method containing specific instructions for downloading from our data provider and returning the desired pandas DataFrame object, which holds our daily market prices, as shown in the following code:

```
def fetch_historical_prices(self):
    import quandl

    # Update your Quandl API key here...
    QUANDL_API_KEY = 'BCzkk3NDWt7H9yjzx-DY'
    quandl.ApiConfig.api_key = QUANDL_API_KEY
    df = quandl.get(self.symbol, start_date=self.start, end_date=self.end)
    return df
```

As this method is specific to Quandl's API, you may rewrite this method to download from your own data provider accordingly.

Also, add the `run()` method inside the `MarketDataSource` class to simulate the streaming prices from our data provider during backtesting:

```
def run(self):
    if self.df is None:
```

```
        self.df = self.fetch_historical_prices()

    total_ticks = len(self.df)
    print('Processing total_ticks:', total_ticks)

    for timestamp, row in self.df.iterrows():
        open_price = row['Open']
        close_price = row['Close']
        volume = row['Volume']

        print(timestamp.date(), 'TICK', self.symbol,
                'open:', open_price,
                'close:', close_price)
        tick_data = TickData(timestamp, self.symbol, open_price,
                             close_price, volume)
        self.market_data.add_tick_data(tick_data)

        if self.tick_event_handler:
            self.tick_event_handler(self.market_data)
```

Notice that the first `if` statement performs a check on the presence of an existing market data before performing the download from our data provider. This allows us to run several simulations on the backtest using the cached data, avoiding the unnecessary download overheads and having our backtests run quicker.

The `for` loop on our `df` market data variable is used to simulate the streaming prices. Each tick data is transformed and formatted as an instance of `TickData` and added to the `market_data` object as the most recently available tick data for that particular symbol. This object is then passed to any tick data event handlers listening to a tick event.

Writing the order class

The `Order` class in the following code represents a single order sent by the strategy to the server. Each order contains a timestamp, the symbol, quantity, and a flag indicating a buy or sell order. In the following examples, we will be using market orders only, and `is_market_order` is expected to be `True`. Other order types, such as limit and stop orders, may be implemented if desired. Once an order is filled, the order is further updated with the filled price, time, and quantity. Write this class as given in the following code:

```
class Order(object):
    def __init__(self, timestamp, symbol,
        qty, is_buy, is_market_order,
        price=0
    ):
```

```
    self.timestamp = timestamp
    self.symbol = symbol
    self.qty = qty
    self.price = price
    self.is_buy = is_buy
    self.is_market_order = is_market_order
    self.is_filled = False
    self.filled_price = 0
    self.filled_time = None
    self.filled_qty = 0
```

Writing a class to keep track of positions

The `Position` class helps us keep track of our current market position and account balance for a traded instrument, and is written as follows:

```
class Position(object):
    def __init__(self, symbol=''):
        self.symbol = symbol
        self.buys = self.sells = self.net = 0
        self.rpnl = 0
        self.position_value = 0
```

The number of units bought, sold, and net are declared as `buys`, `sells`, and `net` variables, respectively. The `rpnl` variable stores the recently realized profits and losses for the symbol. Note that the `position_value` variable starts with a value of zero. When securities are bought, the value of the securities is debited from this account. When securities are sold, the value of the securities is credited into this account.

When an order is filled, an account's position changes. Write a method named `on_position_event()` inside the `Position` class to handle these position events:

```
def on_position_event(self, is_buy, qty, price):
    if is_buy:
        self.buys += qty
    else:
        self.sells += qty

    self.net = self.buys - self.sells
    changed_value = qty * price * (-1 if is_buy else 1)
    self.position_value += changed_value

    if self.net == 0:
        self.rpnl = self.position_value
        self.position_value = 0
```

On a change in our position, we update and keep track of the number of securities bought and sold, as well as the current value of the securities. When the net position is zero, the position is closed out and we obtain the current realized profits and losses.

Whenever a position is open, the value of our securities is influenced by market movements. Having a measure of unrealized profits and losses helps to keep track of the change in market value on every tick movement. Add the following calculate_unrealized_pnl() method inside the Position class:

```python
def calculate_unrealized_pnl(self, price):
    if self.net == 0:
        return 0

    market_value = self.net * price
    upnl = self.position_value + market_value
    return upnl
```

Calling the calculate_unrealized_pnl() method with the current market price gives us the current market value of our position for a particular security.

Writing an abstract strategy class

The Strategy class given in the following code is the base class for all other strategy implementations, and is written as:

```python
from abc import abstractmethod

class Strategy:
    def __init__(self, send_order_event_handler):
        self.send_order_event_handler = send_order_event_handler

    @abstractmethod
    def on_tick_event(self, market_data):
        raise NotImplementedError('Method is required!')

    @abstractmethod
    def on_position_event(self, positions):
        raise NotImplementedError('Method is required!')

    def send_market_order(self, symbol, qty, is_buy, timestamp):
        if self.send_order_event_handler:
            order = Order(
                timestamp,
                symbol,
                qty,
```

```
            is_buy,
            is_market_order=True,
            price=0,
        )
        self.send_order_event_handler(order)
```

The `on_tick_event()` abstract method is called when new market tick data arrives. Child classes would have to implement this abstract method to act upon incoming market prices. The `on_position_event()` abstract method is called whenever there are updates to our positions. Child classes would have to implement this abstract method to act upon incoming position updates.

The `send_market_order()` method is called by child strategy classes to route a market order to the broker. The handler for such an event is stored in the constructor, where the actual implementation is done by the owner of this class in the next section and interfaced directly with a broker API.

Writing a mean-reverting strategy class

In this example, we are implementing a **mean-reverting trading strategy** on the AAPL stock price. Write the `MeanRevertingStrategy` child class that inherits the `Strategy` class from the previous section:

```python
import pandas as pd

class MeanRevertingStrategy(Strategy):
    def __init__(self, symbol, trade_qty,
        send_order_event_handler=None, lookback_intervals=20,
        buy_threshold=-1.5, sell_threshold=1.5
    ):
        super(MeanRevertingStrategy, self).__init__(
            send_order_event_handler)

        self.symbol = symbol
        self.trade_qty = trade_qty
        self.lookback_intervals = lookback_intervals
        self.buy_threshold = buy_threshold
        self.sell_threshold = sell_threshold

        self.prices = pd.DataFrame()
        self.is_long = self.is_short = False
```

In the constructor, we accept parameter values for telling our strategy the security symbol to trade and the number of units for each trade. The `send_order_event_handler` function variable is passed to the parent class to be stored. The `lookback_intervals`, `buy_threshold`, and `sell_threshold` variables are parameters concerned with generating trading signals using mean-reversion calculations.

The `pandas` DataFrame `prices` variable will be used to store incoming prices, and the `is_long` and `is_short` Boolean variables store the current position of this strategy, and only one of them can be `True` at any time. These are assigned in the `on_position_event()` method inside the `MeanRevertingStrategy` class:

```
def on_position_event(self, positions):
    position = positions.get(self.symbol)

    self.is_long = position and position.net > 0
    self.is_short = position and position.net < 0
```

The `on_position_event()` method implements the parent abstract method and gets called on every update of our position.

In addition, implement the `on_tick_event()` abstract method inside the `MeanRevertingStrategy` class:

```
def on_tick_event(self, market_data):
    self.store_prices(market_data)

    if len(self.prices) < self.lookback_intervals:
        return

    self.generate_signals_and_send_order(market_data)
```

On every tick-data event, market prices are stored in the current strategy class to be used for the calculation of trading signals, provided there is sufficient data to so. In this example, we are using daily historical prices with a look-back period of 20 days. In other words, we will be using the mean of the past 20 days' prices to determine a mean reversion. Until there is insufficient data, we simply skip this step.

Add the `store_prices()` method inside the `MeanRevertingStrategy` class:

```
def store_prices(self, market_data):
    timestamp = market_data.get_timestamp(self.symbol)
    close_price = market_data.get_close_price(self.symbol)
    self.prices.loc[timestamp, 'close'] = close_price
```

On each tick event, the `prices` DataFrame store the daily closing price, indexed by a timestamp.

The logic for generating trading signals is given in the `generate_signals_and_send_order()` method inside the `MeanRevertingStrategy` class:

```
def generate_signals_and_send_order(self, market_data):
    signal_value = self.calculate_z_score()
    timestamp = market_data.get_timestamp(self.symbol)

    if self.buy_threshold > signal_value and not self.is_long:
        print(timestamp.date(), 'BUY signal')
        self.send_market_order(
            self.symbol, self.trade_qty, True, timestamp)
    elif self.sell_threshold < signal_value and not self.is_short:
        print(timestamp.date(), 'SELL signal')
        self.send_market_order(
            self.symbol, self.trade_qty, False, timestamp)
```

On each tick event, the **z-score** for the current period is calculated, which we will cover shortly. As soon as the z-score exceeds our buying threshold value, a buy signal is generated. We can either close a short position or enter into a long position by sending a buy market order to our broker. Conversely, when the z-score exceeds our selling threshold value, a sell signal is generated. We can either close a long position or enter into a short position by sending a sell market order to our broker. In our backtest system, orders are executed at the opening of the next day.

Add the `calculate_z_score()` method inside the `MeanRevertingStrategy` class for calculating z-scores on every tick event:

```
def calculate_z_score(self):
    self.prices = self.prices[-self.lookback_intervals:]
    returns = self.prices['close'].pct_change().dropna()
    z_score = ((returns - returns.mean()) / returns.std())[-1]
    return z_score
```

The daily percentage returns of closing prices are z-scored using this formula:

$$\text{z-score} = \frac{x - u}{\sigma}$$

Here, x is the most recent return, μ is the mean of returns, and σ is the standard deviation of returns. A z-score value of 0 indicates that the score is the same as the mean. Take for example a buying threshold value of -1.5. When the z-score falls below -1.5, this indicates a strong buying signal, since z-scores for the following periods are expected to revert to the mean of zero. Similarly, a selling threshold value of 1.5 could indicate a strong selling signal and z-scores are expected to revert to the mean.

Therefore, the goal of this backtesting system aims to find optimal threshold values in maximizing our profits.

Binding our modules with a backtesting engine

After defining all of our core modular components, we are now ready to implement the backtesting engine as the `BacktestEngine` class with the following codes:

```
class BacktestEngine:
    def __init__(self, symbol, trade_qty, start='', end=''):
        self.symbol = symbol
        self.trade_qty = trade_qty
        self.market_data_source = MarketDataSource(
            symbol,
            tick_event_handler=self.on_tick_event,
            start=start, end=end
        )

        self.strategy = None
        self.unfilled_orders = []
        self.positions = dict()
        self.df_rpnl = None
```

Within the backtest engine, we store the symbol and the number of units to trade. An instance of `MarketDataSource` is created with the symbol, together with the start and end dates for defining the timeframe of our dataset. Emitted tick events will be handled by our local `on_tick_event()` method, which we will implement shortly. The `strategy` variable is intended to store an instance of our mean-reverting strategy class. The `unfilled_orders` variable acts as our order book that will store incoming market orders for execution at the next trading day. The `positions` variable is intended to store instances of `Position` objects, indexed by symbol. The `df_rpnl` variable is intended to store our realized profits and losses during backtesting, which we can use to plot at the end of it.

The entry point for running the backtesting engine is the `start()` method given as follows inside the `Backtester` class:

```
def start(self, **kwargs):
    print('Backtest started...')

    self.unfilled_orders = []
    self.positions = dict()
    self.df_rpnl = pd.DataFrame()

    self.strategy = MeanRevertingStrategy(
        self.symbol,
        self.trade_qty,
        send_order_event_handler=self.on_order_received,
        **kwargs
    )
    self.market_data_source.run()

    print('Backtest completed.')
```

A single instance of `Backtester` may be run multiple times by calling the `start()` method. At the start of every run, we initialize the `unfilled_orders`, `positions`, and `df_rpl` variables. A new instance of a strategy class is instantiated with the symbol and number of units to trade, a method named `on_order_received()` for receiving orders triggered from the strategy, as well as any keyword `kwargs` arguments required by the strategy.

Implement the `on_order_received()` method inside the `BacktestEngine` class:

```
def on_order_received(self, order):
    """ Adds an order to the order book """
    print(
        order.timestamp.date(),
        'ORDER',
        'BUY' if order.is_buy else 'SELL',
        order.symbol,
        order.qty
    )
    self.unfilled_orders.append(order)
```

We are notified on the console when an order is generated and added to the order book.

Implement the `on_tick_event()` method inside the `BacktestEngine` class to handle tick events emitted by the market data source:

```python
def on_tick_event(self, market_data):
    self.match_order_book(market_data)
    self.strategy.on_tick_event(market_data)
    self.print_position_status(market_data)
```

The market data source in this example is expected to be the daily historical prices. A tick event received represents a new trading day. At the start of the trading day, we check our order book and match any unfilled orders at the opening by calling the `match_order_book()` method. After which, we pass the latest market data represented by the `market_data` variable to the strategy's tick-event handler to perform trading functions. At the end of the trading day, we print out information on our positions to the console.

Implement the `match_order_book()` and `match_unfilled_orders()` methods inside the `BacktestEngine` class:

```python
def match_order_book(self, market_data):
    if len(self.unfilled_orders) > 0:
        self.unfilled_orders = [
            order for order in self.unfilled_orders
            if self.match_unfilled_orders(order, market_data)
        ]

def match_unfilled_orders(self, order, market_data):
    symbol = order.symbol
    timestamp = market_data.get_timestamp(symbol)

    """ Order is matched and filled """
    if order.is_market_order and timestamp > order.timestamp:
        open_price = market_data.get_open_price(symbol)

        order.is_filled = True
        order.filled_timestamp = timestamp
        order.filled_price = open_price

        self.on_order_filled(
            symbol, order.qty, order.is_buy,
            open_price, timestamp
        )
        return False

    return True
```

On every call of the `match_order_book()` command, a list of pending orders stored in the `unfilled_orders` variable is checked for execution in the market and removed from the list when this operation is successful. The `if` statement in the `match_unfilled_orders()` method verifies that the order is in the correct state and marks the order as filled immediately at the current market-opening price. This would trigger a series of events on the `on_order_filled()` method. Implement this method inside the `BacktestEngine` class:

```
def on_order_filled(self, symbol, qty, is_buy, filled_price, timestamp):
    position = self.get_position(symbol)
    position.on_position_event(is_buy, qty, filled_price)
    self.df_rpnl.loc[timestamp, "rpnl"] = position.rpnl

    self.strategy.on_position_event(self.positions)

    print(
        timestamp.date(),
        'FILLED', "BUY" if is_buy else "SELL",
        qty, symbol, 'at', filled_price
    )
```

As soon as an order is filled, the corresponding position of the traded symbol is required to be updated. The `position` variable contains the retrieved `Position` instance, and a call on its `on_position_event()` command updates its state. The realized profits and losses are calculated and saved to the `pandas` DataFrame `df_rpnl` along with the timestamp. The strategy is also informed of a change in position by calling the `on_position_event()` command. We are notified on the console when such an event occurs.

Add the following `get_position()` method inside the `BacktestEngine` class:

```
def get_position(self, symbol):
    if symbol not in self.positions:
        self.positions[symbol] = Position(symbol)

    return self.positions[symbol]
```

The `get_position()` method is a helper method that simply gets the current `Position` object for a trading symbol. An instance is created if none is found.

The last command call by `on_tick_event()` is `print_position_status()`. Implement this method inside the `BacktestEngine` class:

```
def print_position_status(self, market_data):
    for symbol, position in self.positions.items():
        close_price = market_data.get_close_price(symbol)
```

```
timestamp = market_data.get_timestamp(symbol)

upnl = position.calculate_unrealized_pnl(close_price)

print(
    timestamp.date(),
    'POSITION',
    'value:%.3f' % position.position_value,
    'upnl:%.3f' % upnl,
    'rpnl:%.3f' % position.rpnl
)
```

On every tick event, we print any available position information on the current market value, realized and unrealized profits, and losses to the console.

Running our backtesting engine

With all of the required methods defined inside the `BacktestEngine` class, we may now go ahead and create an instance of this class with the following code:

```
engine = BacktestEngine(
    'WIKI/AAPL', 1,
    start='2015-01-01',
    end='2017-12-31'
)
```

In this example, we are interested in trading one unit of AAPL stock each time, using three years of daily historical data for backtesting from the year 2015 to 2017.

Issue the `start()` command to run the backtesting engine:

```
engine.start(
    lookback_intervals=20,
    buy_threshold=-1.5,
    sell_threshold=1.5
)
```

The `lookback_interval` parameter argument with a value of 20 tells our strategy to use the most recent 20 days' of daily historical prices in calculating z-scores. The `buy_threshold` and `sell_threshold` parameter arguments define the boundary limits at which a buy or sell signal is generated. In this example, a buy threshold value of -1.5 indicates that a long position is desired when the z-score falls below -1.5. Similarly, a sell threshold value of 1.5 indicates that a short position is desired when the z-score rises above 1.5.

When the engine runs, you will see the following output:

```
Backtest started...
Processing total_ticks: 753
2015-01-02 TICK WIKI/AAPL open: 111.39 close: 109.33
...
2015-02-25 TICK WIKI/AAPL open: 131.56 close: 128.79
2015-02-25 BUY signal
2015-02-25 ORDER BUY WIKI/AAPL 1
2015-02-26 TICK WIKI/AAPL open: 128.785 close: 130.415
2015-02-26 FILLED BUY 1 WIKI/AAPL at 128.785
2015-02-26 POSITION value:-128.785 upnl:1.630 rpnl:0.000
2015-02-27 TICK WIKI/AAPL open: 130.0 close: 128.46
```

From the output logs, a buy signal is generated on February 25, 2015 and a market order is added to the order book for execution at the opening of the next trading day on February 26 at USD 128.785. By the end of the trading day, our long position would have an unrealized profit of USD 1.63:

```
...
2015-03-30 TICK WIKI/AAPL open: 124.05 close: 126.37
2015-03-30 SELL signal
2015-03-30 ORDER SELL WIKI/AAPL 1
2015-03-30 POSITION value:-128.785 upnl:-2.415 rpnl:0.000
2015-03-31 TICK WIKI/AAPL open: 126.09 close: 124.43
2015-03-31 FILLED SELL 1 WIKI/AAPL at 126.09
2015-03-31 POSITION value:0.000 upnl:0.000 rpnl:-2.695
...
```

Scrolling further down the logs, you should see that on March 30, 2015 a sell signal is generated, and a sell market order is executed on the next day, March 31, at the price of USD 126.09. This closes our long position and leaves us with a realized loss of USD 2.695.

When the backtest engine finishes, we can plot our strategies realized and profits on to a chart to visualize this trading strategy with the following Python code:

```
%matplotlib inline
import matplotlib.pyplot as plt

engine.df_rpnl.plot(figsize=(12, 8));
```

This gives us the following output:

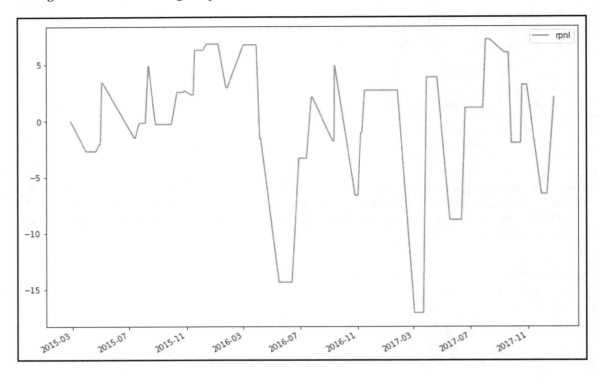

 Note that the realized profits and losses are not complete at the end of the backtest. We may still be holding on to a long or short position with unrealized profits or losses. Be sure to account for this remaining value when evaluating your strategy.

Multiple runs of the backtest engine

Using **fixed strategy parameters**, we were able to have the backtest engine run a single time and visualize its performance. Since the goal of backtesting is to find the optimal strategy parameters for considering in a trading system, we would like our backtest engine to run multiple times on different strategy parameters.

For example, define the list of threshold values that we would like to test in a constant variable named THRESHOLDS:

```
THRESHOLDS = [
    (-0.5,  0.5),
    (-1.5,  1.5),
    (-2.5,  2.0),
    (-1.5,  2.5),
]
```

Each item in the list constant is a tuple of buy and sell threshold values. We can iterate these values with a for loop, calling the engine.start() command and plotting on a graph on every iteration with the following code:

```
%matplotlib inline
import matplotlib.pyplot as plt

fig, axes = plt.subplots(nrows=len(THRESHOLDS)//2,
    ncols=2, figsize=(12, 8))
fig.subplots_adjust(hspace=0.4)
for i, (buy_threshold, sell_threshold) in enumerate(THRESHOLDS):
    engine.start(
        lookback_intervals=20,
        buy_threshold=buy_threshold,
        sell_threshold=sell_threshold
    )
    df_rpnls = engine.df_rpnl
    ax = axes[i // 2, i % 2]
    ax.set_title(
        'B/S thresholds:(%s,%s)' %
        (buy_threshold, sell_threshold)
    )
    df_rpnls.plot(ax=ax)
```

We get the following output:

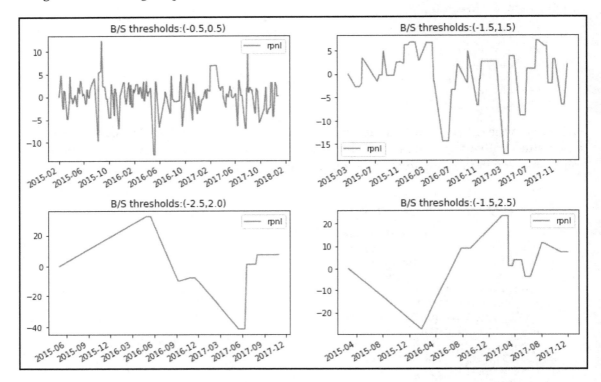

The four plots show the outcomes of the various threshold values being used in our strategy. By varying the strategy parameters, we obtained different profiles of risk and returns. Perhaps you may find better strategy parameters in achieving better results than this!

Improving your backtesting system

In this chapter, we looked at creating a simple backtesting system based on the daily closing prices for a mean-reverting strategy. There are several areas of considerations to make such a backtesting model more realistic. Are historical daily prices sufficient to test our model? Should intra-day limit orders be used instead? Our account value started from zero; how can we reflect our capital requirements accurately? Are we able to borrow shares for shorting?

Since we took an object-oriented approach in creating a backtesting system, how easy would it be to integrate other components in the future? A trading system could accept more than one source of market data. We could also create components that allow us to deploy our system to the production environment.

The list of concerns mentioned are not exhaustive. To guide us in implementing a robust backtesting model, the next section spells out ten considerations in the design of such a system.

Ten considerations for a backtesting model

In the previous section, we performed one replication of a backtest. Our result looks pretty optimistic. However, is this sufficient to deduce that this is a good model? The truth is that backtesting involves a lot of research that merits literature of its own. The following list briefly covers some thoughts that you might want to consider when implementing your backtests.

Resources restricting your model

The resources that are available to your backtesting system limit how well you can implement your backtest. A financial model that generates signals using only the last closing price needs a set of historical data on the closing prices. A trading system that requires reading from the order book requires all levels of the order book data to be available on every tick. This adds up the storage complexity. Other resources, such as exchange data, estimation techniques, and computer resources pose a limitation on the nature of the model that can be used.

Criteria of evaluation of the model

How can we conclude that a model is good? Some factors to consider are Sharpe ratios, hit ratios, average rate of return, VaR statistics, as well as the minimum and maximum drawdown encountered. How can a combination of such factors balance so that a model is usable? How much can the maximum drawdown be tolerated in achieving a high Sharpe ratio?

Estimating the quality of backtest parameters

Using a variety of parameters on a model typically gives us varied results. From multiple models, we can obtain additional sets of data for each model. Can the parameters from the model with the best performance be trustworthy? Using methods such as model averaging can help us correct optimistic estimates.

The model-averaging technique is the average fit for a number of models as opposed to using a single best model.

Be prepared to face model risk

Perhaps after extensive backtesting, you may find yourself with a good-quality model. How long is it going to stay that way? In **model risk**, the market structure or the model parameters may change with time, or a regime change may cause the functional form of your model to change abruptly. By then, you could even be uncertain that your model is correct. A solution that addresses model risk is **model averaging**.

Performance of a backtest with in–sample data

Backtesting helps us perform extensive parameter searches that optimize the results of a model. This exploits the true and the idiosyncratic aspects of the sample data. Also, historical data can never mimic the way that the entire data comes from live markets. These optimized results will always produce an optimistic assessment of the model and the strategy used.

Addressing common pitfalls in backtesting

The most common error made in backtesting is **look-ahead bias**, and it comes in many forms. For example, parameter estimates may be derived from the entire period of the sample data, which constitute using information from the future. Statistical estimates such as these and model selection should be estimated sequentially, which could actually be difficult to do.

Errors in data come in all forms, from hardware, software, and human errors that could occur while routed by data distribution vendors. Listed companies may split, merge, or de-list, resulting in substantial changes to their stock prices. These actions could lead to survivorship bias in our models. Failure to clean data properly will give undue influence to idiosyncratic aspects of data, and thus affect the model parameters.

Survivorship bias is the logical error of concentrating on results that have survived some past selection process. For example, a stock-market index may report a strong performance even in bad times because poor performing stocks are dropped from its component weightage, resulting in an overestimation of past returns.

Failure to use shrinkage estimators or model averaging could report results containing extreme values, making it difficult for comparison and evaluation.

In statistics, a shrinkage estimator is used as an alternative to an ordinary least-squares estimator to produce the smallest mean-squared error. They can be used to shrink raw estimates from the model output toward zero or another fixed constant value.

Have a common-sense idea of your model

Often, common sense could be lacking in our models. We may attempt to explain a trendless variable with a trended variable or infer causation from correlation. Can logarithmic values be used when the context does or does not require it? Let's see in the further section.

Understanding the context for the model

Having a common-sense idea of a model is barely sufficient. A good model takes into account the history, personnel involved, operating constraints, common peculiarities, and all the understanding for the rationale of the model. Are commodity prices following seasonal movements? How was the data gathered? Are the formulas used in the computation of variables reliable? These questions can help us determine the causes, should things go wrong.

Make sure you have the right data

Not many of us have access to tick-level data. Low-resolution tick data may miss out on detailed information. Even tick-level data may be fraught with errors. Using summary statistics, such as the mean, standard errors, maximums, minimums, and correlations, tells us a lot about the nature of the data whether we can really use it, or infer backtest parameter estimates.

When data cleaning is performed, we might ask these questions: what are things to look out for? Are values realistic and logical? How is the missing data coded?

Devise a system of reporting data and results. The use of graphs helps the human eye to visualize patterns that might come across as unexpected. Histograms might reveal an unexpected distribution, or residual plots might show unexpected prediction error patterns. Scatter plots of residualized data may show additional modeling opportunities.

 Residualized data is the difference or *residuals* between the observed values and those of the model.

Data mine your results

From running over several iterations of backtests, the results represent a source of information about your model. Running your model in real-time conditions produces another source of results. By data mining all this wealth of information, we can obtain a data-driven result that can avoid tailoring the model specifications to the sample data. It is recommended to use shrinkage estimators or model averaging when reporting the results.

Discussion of algorithms in backtesting

After taking into consideration the designing of a backtesting model, one or more algorithms may be used to improve the model on a continuous basis. This section briefly covers some of the algorithmic techniques used in areas of backtesting, such as data mining and machine learning.

K-means clustering

The **k-means clustering** algorithm is a method of clustering analysis in data mining. From the backtest results of n observations, the k-means algorithm is designed to classify the data into k clusters based on their relative distance from one another. The center point of each cluster is computed. The objective then is to find the within-cluster sum of squares that gives us a model-averaged point. The model-averaged point indicates the likely average performance of the model, which can be used for further comparison with the performance of other models.

K-nearest neighbors machine learning algorithm

The **k-nearest neighbors** (**KNN**) is a lazy learning technique that does not build any models.

An initial set of backtest model parameters are chosen either by random or best guess.

After analyzing the results of the model, a k number of sets of parameters that is closest to the original set are used for computation in the next step. The model will then take the set of parameters that gives the best results.

The process continues until the terminating condition is reached, thereby always giving the best set of model parameters available.

Classification and regression tree analysis

The **Classification And Regression Tree** (**CART**) analysis contains two decision trees that are used in data mining. The classification tree uses classification rules to classify the outcomes of a model using nodes and branches in the decision tree. The regression tree attempts to assign a real value to the classified outcome. The resulting values are averaged to provide a measure of the quality of the decision.

The 2k factorial design

When designing experiments for backtesting, we can consider the use of **2k factorial design**. Suppose we have two factors, A and B. Each factor behaves as a Boolean value, with values of either +1 or -1. A +1 value indicates a quantitatively high value, while -1 indicates a low value. This gives us a combination of $2^2 = 4$ outcomes. For a 3-factor model, this gives us a combination of $2^3 = 8$ outcomes. The following table illustrates an example with two factors with outcomes W, X, Y, and Z:

	A	B	Replication I
Value	+1	+1	W
Value	+1	-1	X
Value	-1	+1	Y
Value	-1	-1	Z

Note that we are generating one replication of backtest to produce a set of outcomes. Performing additional replications gives us more information. From this data, we can perform a regression and analyze its variance. The objectives of these tests are to determine which factors, A or B, are more influential over another, and what values should be chosen so that the outcomes are either near some desired value, able to achieve a low variance, or minimize the effects of uncontrollable variables.

The genetic algorithm

The **Genetic Algorithm** (**GA**) is a technique where every individual evolves through the process of natural selection in order to optimize a problem. A population of candidate solutions in an optimization problem goes through an iterative process of selection to become parents, undergoing mutation and crossover to produce the next generation of offspring. Over cycles of successive generations, the population evolves toward an optimal solution.

The application of GAs can be applied to a variety of optimizing problems, including backtesting, and is especially useful for solving standard optimizations, discontinuous or non-differentiable problems, or nonlinear outcomes.

Summary

A backtest is a simulation of a model-driven investment strategy's response to historical data. The purpose of performing experiments with backtests is to make discoveries about a process or system and to compute various factors related to either risk or return. The factors are typically used together to find a combination that is predictive of the return.

While working on designing and developing a backtest, it would be helpful to think in terms of the concept of creating video games. In a virtual trading environment, components are needed for the simulation of price feeds, the order-matching engine, the order book management, as well as functions for account and position updates. To achieve these functionalities, we can explore the concept of an event-driven backtesting system.

In this chapter, we designed and implemented a backtesting system that interacts with various components that handle tick data, fetching historical prices from a data provider, handling order and position updates, and simulating a streaming price feed that triggers our strategy to perform mean-reversion calculations. The z-score of each period is evaluated for a trading signal, which leads to the generation of market orders for execution at the opening of the next trading day. We performed a single backtest run as well as multiple runs with varying strategy parameters, plotting the resulting profits and losses to help us visualize the performance of our trading strategy.

Backtesting involves a lot of research that merits literature of its own. In this chapter, we explored ten considerations for designing a backtest model. To help improve our models on a continuous basis, a number of algorithms can be employed in backtesting. We briefly discussed some of these: k-means clustering, k-nearest neighbors, classification and regression trees, 2k factorial design, and genetic algorithms.

In the next chapter, we will learn to perform predictions using machine learning.

Machine Learning for Finance

10

Machine learning is being rapidly adopted for a range of applications in the financial services industry. The adoption of machine learning in financial services has been driven by both supply factors, such as technological advances in data storage, algorithms, and computing infrastructure, and by demand factors, such as profitability needs, competition with other firms, and supervisory and regulatory requirements. Machine learning in finance includes algorithmic trading, portfolio management, insurance underwriting, and fraud detection, just to name a few subject areas.

There are several types of machine learning algorithms, but the two main ones that you will commonly come across in machine learning literature are supervised and unsupervised machine learning. Our discussion in this chapter focuses on supervised learning. Supervised machine learning involves supplying both the input and output data to help the machine predict new input data. Supervised machine learning can be regression-based or classification-based. Regression-based machine learning algorithms predict continuous values, while classification-based machine learning algorithms predict a class or label.

In this chapter, we will be introduced to machine learning, study its concepts and applications in finance, and look at some practical examples for applying machine learning to assist in trading decisions. We will cover the following topics:

- Explore the uses of machine learning in finance
- Supervised and unsupervised machine learning
- Classification-based and regression-based machine learning
- Using scikit-learn for implementing machine learning algorithms
- Applying single-asset regression-based machine learning in predicting prices
- Understanding risk metrics in measuring regression models
- Applying multi-asset regression-based machine learning in predicting returns
- Applying classification-based machine learning in predicting trends
- Understanding risk metrics in measuring classification models

Introduction to machine learning

Before machine learning algorithms became mature, many software application decisions were rule-based, consisting of a bunch of `if` and `else` statements to generate the appropriate response in exchange to some input data. A commonly cited example is a spam filter function in email inboxes. A mailbox may contain blacklisted words defined by a mail server administrator or owner. Incoming emails have their contents scanned against blacklisted words, and should the blacklist condition hold true, the mail is marked as spammed and sent to the `Junk` folder. As the nature of unwanted emails continues to evolve to avoid detection, spam filter mechanisms must also continuously update themselves to keep up with doing a better job. However, with machine learning, spam filters can automatically learn from past email data and, given an incoming email, calculate the possibility of classifying whether the new email is spam or not.

The algorithms behind facial recognition and image detection largely work in the same way. Digital images stored in bits and bytes are collected, analyzed, and classified according to expected responses provided by the owner. This process is known as **training**, using a **supervised learning** approach. The trained data may subsequently be used for predicting the next set of input data as some output response with a certain level of confidence. On the other hand, when the training data does not contain the expected response, the machine learning algorithm is expected to learn from the training data, and this process is called **unsupervised learning**.

Uses of machine learning in finance

Machine learning is increasingly finding its uses in many areas of finance, such as data security, customer service, forecasting, and financial services. A number of use cases leverage big data and **artificial intelligence** (**AI**) as well; they are not exclusive to machine learning. In this section, we will examine some of the ways in which machine learning is transforming the financial sector.

Algorithmic trading

Machine learning algorithms study the statistical properties of the prices of highly correlated assets, measure their predictive power on historical data during backtesting, and forecast prices to within certain accuracy. Machine learning trading algorithms may involve the analysis of the order book, market depth and volume, news releases, earnings calls, or financial statements, where the analysis translates into price movement possibilities and is taken into account for generating trading signals.

Portfolio management

The concept of *robo advisors* has been gaining popularity in recent years to act as automated hedge fund managers. They aid with portfolio construction, optimization, allocation, and rebalancing, and even suggest to clients the instruments to invest in based on their risk tolerance and preferred choice of investment vehicle. These advisories serve as a platform for interacting with a digital financial planner, providing financial advice and portfolio management.

Supervisory and regulatory functions

Financial institutions and regulators are adopting the use of AI and machine learning to analyze, identify, and flag suspicious transactions that warrant further investigation. Supervisors such as the **Securities and Exchange Commission** (**SEC**) take a data-driven approach and employ AI, machine learning, and natural language processing to identify behavior that warrants enforcement. Worldwide, central authorities are developing machine learning capabilities in regulatory functions.

Insurance and loan underwriting

Insurance companies actively use AI and machine learning to augment some insurance sector functions, improve pricing and marketing of insurance products, and to reduce claims processing times and operational costs. In loan underwriting, many data points of a single consumer, such as age, income, and credit score, are compared against a database of candidates in building credit risk profiles, determining credit scores, and calculating the possibility of loan defaults. Such data relies on transaction and payment history from financial institutions. However, lenders are increasingly turning to social media activities, mobile phone usage, and messaging activities to capture a more holistic view of creditworthiness, speed up lending decisions, limit incremental risk, and improve the rating accuracy of loans.

News sentiment analysis

Natural language processing, a subset of machine learning, may be used to analyze alternative data, financial statements, news announcements, and even Twitter feeds, in creating investment sentiment indicators used by hedge funds, high-frequency trading firms, social trading, and investment platforms for analyzing markets in real time. Politicians' speeches, or important new releases, such as those made by central banks, are also being analyzed in real time, where each and every word is being scrutinized and calculated to predict which asset prices could move and by how much. Machine learning will not only understand the movement of stock prices and trades, but also understand social media feeds, news trends, and other data sources.

Machine learning beyond finance

Machine learning is increasingly being employed in areas of facial recognition, voice recognition, biometrics, trade settlement, chatbots, sales recommendations, content creation, and more. As machine learning algorithms improve and their rate of adoption picks up, the list of use cases becomes even longer.

Let's begin our journey in machine learning by understanding some of the terminology you will come across in the machine learning literature.

Supervised and unsupervised learning

There are many types of machine learning algorithms, but the two main ones that you will commonly come across are supervised and unsupervised machine learning.

Supervised learning

Supervised learning predicts a certain output from given inputs. These pairings of input to output data are known as **training data**. The quality of the prediction entirely depends on the training data; incorrect training data reduces the effectiveness of the machine learning model. An example is a dataset of transactions with labels identifying which ones are fraudulent, and which are not. A model can then be built to predict whether a new transaction will be fraudulent.

Some common algorithms in supervised learning are logistic regression, the support vector machine, and random forests.

Unsupervised learning

Unsupervised learning builds a model based on given input data that does not contain labels, but instead is asked to detect patterns in the data. This may involve identifying clusters of observations with similar underlying characteristics. Unsupervised learning aims to make accurate predictions to new, never-before-seen data.

For example, an unsupervised learning model may price illiquid securities by looking for a cluster of securities with similar characteristics. Common unsupervised learning algorithms include k-means clustering, principal component analysis, and autoencoders.

Classification and regression in supervised machine learning

There are two major types of supervised machine algorithms, mainly classification and regression. Classification machine learning models attempt to predict and classify responses from a list of predefined possibilities. These predefined possibilities may be binary classification (such as a *Yes* or *No* response to a question: *Is this email spam?*) or multi-class classification.

Regression machine learning models attempt to predict continuous output values. For example, predicting housing prices or the temperature expects a continuous range of output values. Common forms of regressions are **ordinary least squares** (**OLS**) regression, LASSO regression, ridge regression, and elastic net regularization.

Overfitting and underfitting models

Poor performance in machine learning models can be caused by overfitting or underfitting. An overfitted machine learning model is one that is trained too well with the training data such that it leads to negative performance on new data. This occurs when the training data is fitted to every minor variation, including noise and random fluctuations. Unsupervised learning algorithms are highly susceptible to overfitting, since the model learns from every piece of data, both good and bad.

An underfitted machine learning model gives poor accuracy of prediction. It may be caused by too little training data being available to build an accurate model, or that the data is not suitable for extracting its underlying trends. Underfitting models are easy to detect since they give consistently poor performance. To improve such models, provide more training data or use another machine learning algorithm.

Feature engineering

A feature is an attribute of the data that defines its characteristic. By using domain knowledge of the data, features can be created to help machine learning algorithms increase their predictive performance. This can be as simple as grouping or bucketing related parts of the existing data to form defining features. Even removing unwanted features is also feature engineering.

As an example, suppose we have the following time series price data that looks like this:

No.	Date and Time	Price	Price Action
1	2019-01-02 09:00:01	55.00	UP
2	2019-01-02 10:03:42	45.00	DOWN
3	2019-01-02 10:31:23	48.00	UP
4	2019-01-02 11:14:02	33.00	DOWN

Grouping the time series into buckets by the hour of the day and taking the last price action in each bucket, we end up with a feature like this:

No.	Hour of Day	Last Price Action
1	9	UP
2	10	UP
3	11	DOWN

The process of feature engineering involves these four steps:

1. Brainstorming features to include in the training model
2. Creating those features
3. Checking how the features work with the model
4. Repeating from step 1 until the features work perfectly

There are absolutely no hard and fast rules when it comes to what constitutes creating features. Feature engineering is considered more of an art than a science.

Scikit-learn for machine learning

Scikit-learn is a Python library designed for scientific computing and contains a number of state-of-the-art machine learning algorithms for classification, regression, clustering, dimensionality reduction, model selection, and preprocessing. Its name is derived from the SciPy Toolkit, which is an extension of the SciPy module. Comprehensive documentation on scikit-learn can be found at `https://scikit-learn.org`.

 SciPy is a collection of Python modules for scientific computing, containing a number of core packages, such as NumPy, Matplotlib, IPython, and others.

In this chapter, we will be using scikit-learn's machine learning algorithms to predict securities movements. Scikit-learn require a working installation of NumPy and SciPy. Install scikit-learn with the `pip` package manager by using the following command:

```
pip install scikit-learn
```

Predicting prices with a single-asset regression model

Pairs trading is a common statistical arbitrage trading strategy employed by traders using a pair of co-integrated and highly positively correlated assets, though negatively correlated pairs can also be considered.

In this section, we will use machine learning to train regression-based models using the historical prices of a pair of securities that might be used in pairs trading. Given the current price of one security for a particular day, we predict the other security's price on a daily basis. The following examples uses the historical daily prices of **Goldman Sachs** (**GS**) and **J.P. Morgan** (**JPM**) traded on the **New York Stock Exchange** (**NYSE**). We will be predicting prices of JPM's stock price for the year 2018.

Linear regression by OLS

Let's begin our investigation of regression-based machine learning with a simple linear regression model. A straight line is in the following form:

$$\hat{y} = ax + c$$

This attempts to fit the data by OLS:

- a is the slope or coefficient
- c is the value of the y-intercept
- x is the input dataset
- $\hat{y}$ is the predicted value from the straight line

The coefficients and intercept are determined by minimizing the cost function:

$$minimize \sum_{i=0}^{n-1}(y_i - \hat{y}_i)^2$$

y is the dataset of observed actual values used in performing a straight-line fit. In other words, we are performing a least sum of squared errors in finding the coefficients a and c, from which we can predict the current period.

Before developing a model, let's download and prepare the required datasets.

Preparing the independent and target variables

Let's obtain the datasets of GS and JPM prices from Alpha Vantage with the following code:

```
In [ ]:
    from alpha_vantage.timeseries import TimeSeries

    # Update your Alpha Vantage API key here...
    ALPHA_VANTAGE_API_KEY = 'PZ2ISG9CYY379KLI'

    ts = TimeSeries(key=ALPHA_VANTAGE_API_KEY, output_format='pandas')
    df_jpm, meta_data = ts.get_daily_adjusted(
        symbol='JPM', outputsize='full')
    df_gs, meta_data = ts.get_daily_adjusted(
        symbol='GS', outputsize='full')
```

The `pandas` DataFrame objects `df_jpm` and `df_gs` contain the downloaded prices of JPM and GS respectively. We will be extracting the adjusted closing prices from the fifth column of each dataset.

Let's prepare our independent variables with the following code:

```
In [ ]:
    import pandas as pd

    df_x = pd.DataFrame({'GS': df_gs['5. adjusted close']})
```

The adjusted closing prices of GS are extracted to a new DataFrame object, df_x. Next, obtain our target variables with the following code:

```
In [ ]:
    jpm_prices = df_jpm['5. adjusted close']
```

The adjusted closing prices of JPM are extracted to the jpm_prices variable as a pandas Series object. Having prepared our datasets for use in modeling, let's proceed to develop the linear regression model.

Writing the linear regression model

We will create a class for using a linear regression model to fit and predict values. This class also serves as a base class for implementing other models in this chapter. The following steps illustrates this process.

1. Declare a class named LinearRegressionModel as follows:

```
from sklearn.linear_model import LinearRegression

class LinearRegressionModel(object):
    def __init__(self):
        self.df_result = pd.DataFrame(columns=['Actual',
'Predicted'])

    def get_model(self):
        return LinearRegression(fit_intercept=False)

    def get_prices_since(self, df, date_since, lookback):
        index = df.index.get_loc(date_since)
        return df.iloc[index-lookback:index]
```

In the constructor of our new class, we declare a pandas DataFrame called df_result to store the actual and predicted values for plotting on a chart later on. The get_model() method returns an instance of the LinearRegression class in the sklearn.linear_model module for fitting and predicting the data. The set_intercept parameter is set to True as the data is not centered (around 0 on the *x*- and *y*-axes, that is).

> More information about the LinearRegression of scikit-learn can be found at https://scikit-learn.org/stable/modules/generated/ sklearn.linear_model.LinearRegression.html.

The get_prices_since() method slices a subset of the supplied dataset with the iloc command, from the given date index date_since and up to a number of earlier periods defined by the lookback value.

2. Add a method named learn() into the LinearRegressionModel class, as follows:

```
def learn(self, df, ys, start_date, end_date, lookback_period=20):
    model = self.get_model()

    for date in df[start_date:end_date].index:
        # Fit the model
        x = self.get_prices_since(df, date, lookback_period)
        y = self.get_prices_since(ys, date, lookback_period)
        model.fit(x, y.ravel())

        # Predict the current period
        x_current = df.loc[date].values
        [y_pred] = model.predict([x_current])

        # Store predictions
        new_index = pd.to_datetime(date, format='%Y-%m-%d')
        y_actual = ys.loc[date]
        self.df_result.loc[new_index] = [y_actual, y_pred]
```

The learn() method serves as the entry point for running the model. It accepts the df and ys parameters as our independent and target variables, start_date and end_date as strings corresponding to the index of the dataset for the period we will be predicting, and the lookback_period parameter as the number of historical data points used for fitting the model in the current period.

The `for` loop simulates a backtest on a daily basis. The call to `get_prices_since()` fetches a subset of the dataset for fitting the model on the *x*- and *y*-axes with the `fit()` command. The `ravel()` command transforms the given `pandas` Series object into a flattened list of target values for fitting the model.

The `x_current` variable represents independent variable values for the specified date, fed into the `predict()` method. The predicted output is a `list` object, from which we extract the first value. Both the actual and predicted values are saved to the `df_result` DataFrame, indexed by the current date as a `pandas` object.

3. Let's instantiate this class and run our machine learning model by issuing the following commands:

```
In [ ]:
    linear_reg_model = LinearRegressionModel()
    linear_reg_model.learn(df_x, jpm_prices, start_date='2018',
                        end_date='2019', lookback_period=20)
```

In the `learn()` command, we provided our prepared datasets, `df_x` and `jpm_prices`, and specified the prediction for the year of 2018. For this example, we assumed there are 20 trading days in a month. Using a `lookback_period` value of `20`, we are using a past month's prices to fit our model for prediction daily.

4. Let's retrieve the resulting `df_result` DataFrame from the model and plot both the actual and predicted values:

```
In [ ]:
    %matplotlib inline

    linear_reg_model.df_result.plot(
        title='JPM prediction by OLS',
        style=['-', '--'], figsize=(12,8));
```

In the `style` parameter, we specified that actual values are to be drawn as a solid line, and predicted values drawn as dotted lines. This gives us the following graph:

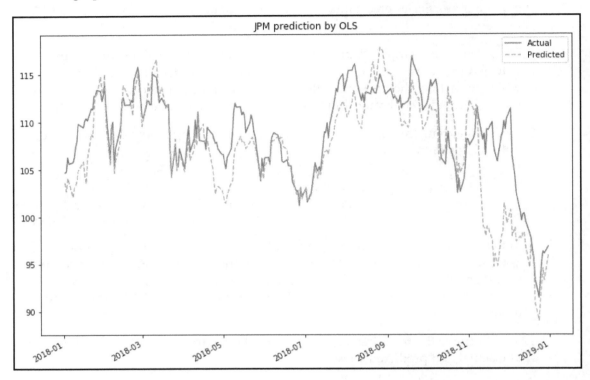

The chart shows our predicted results trailing closely behind the actual values up to a certain extent. How well does our model actually perform? In the next section, we will discuss several common risk metrics for measuring regression-based models.

Risk metrics for measuring prediction performance

The `sklearn.metrics` module implements several regression metrics for measuring prediction performance. We will discuss the mean absolute error, the mean squared error, the explained variance score, and the R^2 score in subsequent sections.

Mean absolute error as a risk metric

The **mean absolute error** (**MAE**) is a risk metric that measures the average absolute prediction error and can be written as follows:

$$MAE(y, \hat{y}) = \frac{1}{n} \sum_{i=0}^{n-1} |y_i - \hat{y}_i|$$

Here, y and $\hat{y}$ are the actual and predicted lists of values, respectively, with the same length, n. $\hat{y}_i$ and y_i are the predicted and actual values, respectively, at the index i. Taking the absolute values of errors means that our output results in a positive decimal value. Low values of MAE are highly desired. A perfect score of 0 implies that our prediction powers are exactly aligned with actual values, since there are no differences between the two.

Obtain the MAE value of our predictions using the `mean_absolute_error` function of the `sklearn.metrics` module with the following code:

```
In [ ]:
    from sklearn.metrics import mean_absolute_error

    actual = linear_reg_model.df_result['Actual']
    predicted = linear_reg_model.df_result['Predicted']

    mae = mean_absolute_error(actual, predicted)
    print('mean absolute error:', mae)
Out[ ]:
    mean absolute error: 2.4581692107823367
```

The MAE of our linear regression model is 2.458.

Mean squared error as a risk metric

Like the MAE, the **mean squared error** (**MSE**) is a risk metric that measures the average of the squares of the prediction errors and can be written as follows:

$$MSE(y, \hat{y}) = \frac{1}{n} \sum_{i=0}^{n-1} (y_i - \hat{y}_i)^2$$

Squaring the errors means that values of MSE are always positive, and low values of MSE are highly desired. A perfect MSE score of 0 implies that our prediction powers are exactly aligned with actual values, and that the squares of such differences are negligible. While the application of both the MSE and MAE helps determine the strength of our model's predictive powers, MSE triumphs over MAE by penalizing errors that are farther away from the mean. Squaring the errors places a heavier bias on the risk metrics.

Obtain the MSE value of our predictions using the `mean_squared_error` function of the `sklearn.metrics` module with the following code:

```
In [ ]:
    from sklearn.metrics import mean_squared_error
    mse = mean_squared_error(actual, predicted)
    print('mean squared error:', mse)
Out[ ]:
    mean squared error: 12.156835196436589
```

The MSE of our linear regression model is 12.156.

Explained variance score as a risk metric

The explained variance score explains the dispersion of errors of a given dataset, and the formula is written as follows:

$$explained\ variance(y, \hat{y}) = 1 - \frac{Var(y - \hat{y})}{Var(y)}$$

Here, $Var(y - \hat{y})$ and $Var(y)$ is the variance of prediction errors and actual values respectively. Scores close to 1.0 are highly desired, indicating better squares of standard deviations of errors.

Obtain the explained variance score of our predictions using the `explained_variance_score` function of the `sklearn.metrics` module with the following code:

```
In [ ]:
    from sklearn.metrics import explained_variance_score
    eva = explained_variance_score(actual, predicted)
    print('explained variance score:', eva)
Out[ ]:
    explained variance score: 0.5332235487812286
```

The explained variance score of our linear regression model is 0.533.

R² as a risk metric

The R^2 score is also known as the **coefficient of determination,** and it measures how well future samples are likely to be predicted by the model. It is written as follows:

$$R^2(y, \hat{y}) = 1 - \frac{\sum_{i=0}^{n-1}(y_i - \hat{y})^2}{\sum_{i=0}^{n-1}(y_i - \bar{y})^2}$$

Here, $\bar{y}$ is the mean of actual values and can be written as follows:

$$\bar{y} = \frac{1}{n}\sum_{i=0}^{n-1} y_i$$

R^2 scores ranges from negative values to 1.0. A perfect R^2 score of 1 implies that there is no error in the regression analysis, while a score of 0 indicates that the model always predicts the mean of target values. A negative R^2 score indicates that the prediction performs below average.

Obtain the R^2 score of our predictions using the `r2_score` function of the `sklearn.metrics` module with the following code:

```
In [ ]:
    from sklearn.metrics import r2_score
    r2 = r2_score(actual, predicted)
    print('r2 score:', r2)
Out[ ]:
    r2 score: 0.41668246393290576
```

The R^2 of our linear regression model is 0.4167. This implies that 41.67% of the variability of the target variables have been accounted for.

Ridge regression

The ridge regression, or L2 regularization, addresses some of the problems of OLS regression by penalizing the sum of squares of the model coefficients. The cost function for the ridge regression can be written as follows:

$$minimize \sum_{i=1}^{n}(y_i - \hat{y})^2 + \alpha \sum_{j=0}^{m} b_j^2$$

Here, the α parameter is expected to be a positive value that controls the amount of shrinkage. Larger values of alpha give greater shrinkage, making the coefficients more robust to collinearity.

The `Ridge` class of the `sklearn.linear_model` module implements ridge regression. To implement this model, create a class named `RidgeRegressionModel` that extends the `LinearRegressionModel` class, and run the following code:

```
In [ ]:
    from sklearn.linear_model import Ridge

    class RidgeRegressionModel(LinearRegressionModel):
        def get_model(self):
            return Ridge(alpha=.5)

    ridge_reg_model = RidgeRegressionModel()
    ridge_reg_model.learn(df_x, jpm_prices, start_date='2018',
                    end_date='2019', lookback_period=20)
```

In the new class, the `get_model()` method is overridden to return the ridge regression model of scikit-learn while reusing the other methods in the parent class. The `alpha` value is set to 0.5, and the rest of the model parameters are left as defaults. The `ridge_reg_model` variable represents an instance of our ridge regression model, and the `learn()` command is run with the usual parameters values.

Create a function called `print_regression_metrics()` to print the various risk metrics covered earlier:

```
In [ ]:
    from sklearn.metrics import (
        accuracy_score, mean_absolute_error,
        explained_variance_score, r2_score
    )
    def print_regression_metrics(df_result):
        actual = list(df_result['Actual'])
        predicted = list(df_result['Predicted'])
        print('mean_absolute_error:',
            mean_absolute_error(actual, predicted))
        print('mean_squared_error:', mean_squared_error(actual, predicted))
        print('explained_variance_score:',
            explained_variance_score(actual, predicted))
        print('r2_score:', r2_score(actual, predicted))
```

Pass the `df_result` variable to this function and display the risk metrics to the console:

```
In [ ]:
    print_regression_metrics(ridge_reg_model.df_result)
Out[ ]:
    mean_absolute_error: 1.5894879428144535
    mean_squared_error: 4.519795633665941
    explained_variance_score: 0.7954229624785825
    r2_score: 0.7831280913202121
```

Both mean error scores of the ridge regression model are lower than the linear regression model and are closer to zero. The explained variance score and the R² score are higher than the linear regression model and are closer to 1. This indicates that our ridge regression model is doing a better job of prediction than the linear regression model. Besides having better performance, ridge regression computations are less costly than the original linear regression model.

Other regression models

The `sklearn.linear_model` module contains various regression models that we can consider implementing in our model. The remaining sections briefly describe them. A full list of linear models is available at `https://scikit-learn.org/stable/modules/classes.html#module-sklearn.linear_model`.

Lasso regression

Similar to ridge regression, **Least Absolute Shrinkage and Selection Operator (LASSO)** regression is also another form of regularization that involves penalizing the sum of absolute values of regression coefficients. It uses the L1 regularization technique. The cost function for the LASSO regression can be written as follows:

$$minimize \sum_{i=1}^{n}(y_i - \hat{y})^2 + \alpha \sum_{j=0}^{m}|b_j|$$

Like ridge regression, the alpha parameter α controls the strength of the penalty. However, for geometric reasons, LASSO regression produces different results than ridge regression since it forces a majority of the coefficients to be set to zero. It is better suited for estimating sparse coefficients and models with fewer parameter values.

The `Lasso` class of `sklearn.linear_model` implements LASSO regression.

Elastic net

Elastic net is another regularized regression method that combines the L1 and L2 penalties of the LASSO and ridge regression methods. The cost function for elastic net can be written as follows:

$$minimize \frac{1}{2n} \sum_{i=1}^{n} (y_i - \hat{y})^2 + \alpha_1 \sum_{j=0}^{m} |b_j| + \alpha_2 \sum_{j=0}^{m} b_j^2$$

The alpha values are explained here:

$$\alpha_1 = alpha * \text{l1\_ratio}$$

$$\alpha_2 = 0.5 * alpha * (1 - \text{l1\_ratio})$$

Here, `alpha` and `l1_ratio` are parameters of the `ElasticNet` function. When `alpha` is zero, the cost function is equivalent to an OLS. When `l1_ratio` is zero, the penalty is a ridge or L2 penalty. When `l1_ratio` is 1, the penalty is a LASSO or L1 penalty. When `l1_ratio` is between 0 and 1, the penalty is a combination of L1 and L2.

The `ElasticNet` class of `sklearn.linear_model` implements elastic net regression.

Conclusion

We used a single-asset, trend-following momentum strategy by regression to predict the prices of JPM using GS, with the assumption that the pair is cointegrated and highly correlated. We can also consider cross-asset momentum to obtain better results from diversification. The next section explores multi-asset regression for predicting security returns.

Predicting returns with a cross-asset momentum model

In this section, we will create a cross-asset momentum model by having the prices of four diversified assets predict the returns of JPM on a daily basis for the year of 2018. The prior 1-month, 3-month, 6-month, and 1-year of lagged returns of the S&P 500 stock index, 10-year treasury bond index, US dollar index, and gold prices will be used for fitting our model. This gives us a total of 16 features. Let's begin by preparing our datasets for developing our models.

Preparing the independent variables

We will use Alpha Vantage again as our data provider. As this free service does not provide all of the dataset required for our investigation, we shall consider other close assets as a proxy. The ticker symbol for the S&P 500 stock index is SPX. We will use the SPDR Gold Trust (ticker symbol: GLD) to denote a share of the gold bullion as a proxy for gold prices. The Invesco DB US Dollar Index Bullish Fund (ticker symbol: UUP) will proxy the US dollar index. The iShares 7-10 Year Treasury Bond ETF (ticker symbol: IEF) will proxy the 10-year Treasury Bond Index. Run the following code to download our datasets:

```
In [ ]:
    df_spx, meta_data = ts.get_daily_adjusted(
        symbol='SPX', outputsize='full')
    df_gld, meta_data = ts.get_daily_adjusted(
        symbol='GLD', outputsize='full')
    df_dxy, dxy_meta_data = ts.get_daily_adjusted(
        symbol='UUP', outputsize='full')
    df_ief, meta_data = ts.get_daily_adjusted(
        symbol='IEF', outputsize='full')
```

The `ts` variable is the `TimeSeries` object of Alpha Vantage created in the previous section. Combine the adjusted closing prices into a single `pandas` DataFrame named `df_assets` with the following codes and remove empty values with the `dropna()` command:

```
In [ ]:
    import pandas as pd

    df_assets = pd.DataFrame({
        'SPX': df_spx['5. adjusted close'],
        'GLD': df_gld['5. adjusted close'],
        'UUP': df_dxy['5. adjusted close'],
        'IEF': df_ief['5. adjusted close'],
```

```
    }).dropna()
```

Calculate the lagged percentage returns of our `df_assets` dataset with the following code:

```
IN [ ]:
    df_assets_1m = df_assets.pct_change(periods=20)
    df_assets_1m.columns = ['%s_1m'%col for col in df_assets.columns]

    df_assets_3m = df_assets.pct_change(periods=60)
    df_assets_3m.columns = ['%s_3m'%col for col in df_assets.columns]

    df_assets_6m = df_assets.pct_change(periods=120)
    df_assets_6m.columns = ['%s_6m'%col for col in df_assets.columns]

    df_assets_12m = df_assets.pct_change(periods=240)
    df_assets_12m.columns = ['%s_12m'%col for col in df_assets.columns]
```

In the `pct_change()` command, the `periods` parameter specifies the number of periods to shift. We assumed 20 trading days in a month when calculating the lagged returns. Combine the four `pandas` DataFrame objects into a single DataFrame with the `join()` command:

```
In [ ]:
    df_lagged = df_assets_1m.join(df_assets_3m)\
        .join(df_assets_6m)\
        .join(df_assets_12m)\
        .dropna()
```

Use the `info()` command to view its properties:

```
In [ ]:
    df_lagged.info()
Out[ ]:
    <class 'pandas.core.frame.DataFrame'>
    Index: 2791 entries, 2008-02-12 to 2019-03-14
    Data columns (total 16 columns):
    . . .
```

The output is truncated, but you can see 16 features as our independent variables spanning the years 2008 to 2019. Let's continue to obtain the dataset for our target variables.

Preparing the target variables

The closing prices of JPM having been downloaded to the `pandas` Series object `jpm_prices` earlier, simply calculate the actual percentage returns with the following code:

```
In [ ]:
    y = jpm_prices.pct_change().dropna()
```

We obtain a `pandas` Series object as our target variable `y`.

A multi-asset linear regression model

In the previous section, we used a single asset with the prices of GS for fitting our linear regression model. This same model, `LinearRegressionModel`, accommodates multiple assets. Run the following commands to create an instance of this model and use our new datasets:

```
In [ ]:
    multi_linear_model = LinearRegressionModel()
    multi_linear_model.learn(df_lagged, y, start_date='2018',
                             end_date='2019', lookback_period=10)
```

In the linear regression model instance, `multi_linear_model`, the `learn()` command is supplied with the `df_lagged` dataset with 16 features and `y` as the percentage changes of JPM. The `lookback_period` value is reduced in consideration of the limited lagged returns data available. Let's plot the actual versus predicted percentage changes of JPM:

```
In [ ]:
    multi_linear_model.df_result.plot(
        title='JPM actual versus predicted percentage returns',
        style=['-', '--'], figsize=(12,8));
```

This would give us the following graph in which the solid lines show the actual percentage returns of JPM, while the dotted lines show the predicted percentage returns:

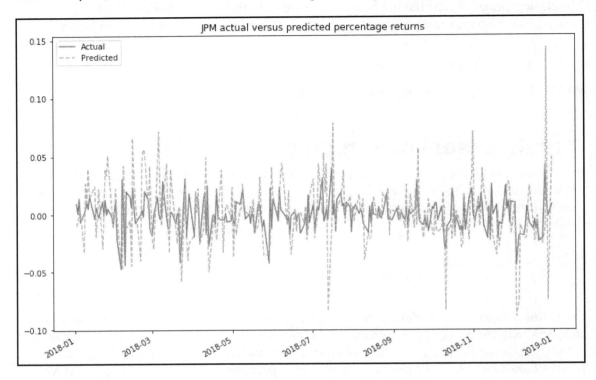

How well did our model perform? Let's run the same performance metrics in the `print_regression_metrics()` function defined in the previous section:

```
In [ ]:
    print_regression_metrics(multi_linear_model.df_result)
Out [ ]:
    mean_absolute_error: 0.01952328066607389
    mean_squared_error: 0.0007225502867195044
    explained_variance_score: -2.729798588246765
    r2_score: -2.738404583097052
```

The explained variance score and R^2 scores are in the negative range, suggesting that the model performs below average. Can we perform better? Let's explore more complex tree models used in regression.

An ensemble of decision trees

Decision trees are widely used models for classification and regression tasks, much like a binary tree, where each node represents a question leading to a yes-no answer for traversing the respective left and right nodes. The goal is to get to the right answer by asking as few questions as possible.

 A paper describing deep neural decision trees can be found at `https:// arxiv.org/pdf/1806.06988.pdf`.

Traversing deep down decision trees can quickly lead to overfitting of the given data, rather than inferring the overall properties of the distributions from which they are drawn. To address this issue of overfitting, the data can be split into subsets and train on different trees, each on a subset. This way, we end up with an ensemble of different decision tree models. When random subsets of the samples are drawn with replacements for prediction, this method is called **bagging** or **bootstrap aggregation**. We may or may not get consistent results across these models, but the final model obtained by averaging the bootstrapped models yields better results than using a single decision tree. Using an ensemble of randomized decision trees is known as **random forests**.

Let's visit some decision tree models in scikit-learn that we may consider implementing in our multi-asset regression model.

Bagging regressor

The `BaggingRegressor` class of `sklearn.ensemble` implements the bagging regressor. We can see how a bagging regressor works for multi-asset predictions of the percentage returns of JPM. The following code illustrates this:

```
In [ ]:
    from sklearn.ensemble import BaggingRegressor

    class BaggingRegressorModel(LinearRegressionModel):
        def get_model(self):
            return BaggingRegressor(n_estimators=20, random_state=0)
In [ ]:
    bagging = BaggingRegressorModel()
    bagging.learn(df_lagged, y, start_date='2018',
                  end_date='2019', lookback_period=10)
```

We created a class named `BaggingRegressorModel` that extends `LinearRegressionModel`, and the `get_model()` method is overridden to return the bagging regressor. The `n_estimators` parameter specifies `20` base estimators or decision trees in the ensemble, with the `random_state` parameter as a seed of `0` used by the random number generator. The rest of the parameters are default values. We run this model with the same dataset.

Run the same performance metrics and see how our model performs:

```
In [ ]:
    print_regression_metrics(bagging.df_result)
Out[ ]:
    mean_absolute_error: 0.0114699264723
    mean_squared_error: 0.000246352185742
    explained_variance_score: -0.272260304849
    r2_score: -0.274602137956
```

The MAE and MSE values indicate that an ensemble of decision trees produces fewer prediction errors than the simple linear regression model. Also, though the explained variance score and the R^2 scores are negative, it indicates a better variance of data towards the mean than is offered by the simple linear regression model.

Gradient tree boosting regression model

Gradient tree boosting, or simply gradient boosting, is a technique of improving or boosting the performance of weak learners using a gradient descent procedure to minimize the loss function. Tree models, usually decision trees, are added one at a time and build the model in a stage-wise fashion, while leaving the existing trees in the model unchanged. Since gradient boosting is a greedy algorithm, it can overfit a training dataset quickly. However, it can benefit from regularization methods that penalize various parts of the algorithm and reduce overfitting to improve its performance.

The `sklearn.ensemble` module provides a gradient-boosting regressor called `GradientBoostingRegressor`.

Random forests regression

Random forests consist of multiple decision trees each based on a random sub-sample of the training data and uses averaging to improve the predictive accuracy and to control overfitting. Selection by random inadvertently introduces some form of bias. However, due to averaging, it variance also decreases, helping to compensate for the increase in bias, and is considered to yield an overall better model.

The `sklearn.ensemble` module provides a random forest regressor called `RandomForestRegressor`.

More ensemble models

The `sklearn.ensemble` module contains various other ensemble regressors, as well as classifier models. More information can be found at `https://scikit-learn.org/stable/modules/classes.html#module-sklearn.ensemble`.

Predicting trends with classification-based machine learning

Classification-based machine learning is a supervised machine learning approach in which a model learns from given input data and classifies it according to new observations. Classification may be bi-class, such as identifying whether an option should be exercised or not, or multi-class, such as the direction of a price change, which can be either up, down, or unchanging.

In this section, we will look again at creating cross-asset momentum models by having the prices of four diversified assets predict the daily trend of JPM on a daily basis for the year of 2018. The prior 1-month and 3-month lagged returns of the S&P 500 stock index, the 10-year treasury bond index, the US dollar index, and gold prices will be used to fit the model for prediction. Our target variables consist of Boolean indicators, where a `True` value indicates an increase or no-change from the previous trading day's closing price, and a `False` value indicates a decrease.

Let's begin by preparing the dataset for our models.

Preparing the target variables

We have already downloaded the JPM dataset to the `pandas` DataFrame, `df_jpm`, in a previous section, and the `y` variable contains the daily percentage change of JPM. Convert these values to labels with the following code:

```
In [ ]:
    import numpy as np
    y_direction = y >= 0
    y_direction.head(3)
Out[ ]:
```

```
date
1998-01-05      True
1998-01-06      False
1998-01-07      True
Name: 5. adjusted close, dtype: bool
```

Using the `head()` command, we can see that the `y_direction` variable becomes a `pandas` Series object of Boolean values. A percentage change of zero or more classifies the value with a `True` label, and `False` otherwise. Let's extract unique values with the `unique()` command as column names for use later on:

```
In [ ]:
    flags = list(y_direction.unique())
    flags.sort()
    print(flags)
Out[ ]:
    [False, True]
```

The column names are extracted to a variable called `flags`. With our target variables ready, let's continue to obtain our independent multi-asset variables.

Preparing the dataset of multiple assets as input variables

We will be reusing the `pandas` DataFrame variables, `df_assets_1m` and `df_assets_3m`, from the previous section containing the lagged 1-month and 3-month percentage returns of the four assets and combine them into a single variable, `df_input`, with the following code:

```
In [ ]:
    df_input = df_assets_1m.join(df_assets_3m).dropna()
```

Use the `info()` command to view its properties:

```
In [ ]:
    df_input.info()
Out[ ]:
    <class 'pandas.core.frame.DataFrame'>
    Index: 2971 entries, 2007-05-25 to 2019-03-14
    Data columns (total 8 columns):
    ...
```

The output is truncated, but you can see we have eight features as our independent variables spanning the years 2007 to 2019. With our input and target variables created, let's explore the various classifiers available in scikit-learn for modeling.

Logistic regression

Despite its name, logistic regression is actually a linear model used for classification. It uses a logistic function, also known as a **sigmoid** function, to model the probabilities describing the possible outcomes of a single trial. A logistic function helps to map any real-valued number to a value between 0 and 1. A standard logistic function is written as follows:

$$\hat{y} = \frac{1}{1 + e^{-x}}$$

e is the base of the natural logarithm, and x is the X-value of the sigmoid's midpoint. $\hat{y}$ is the predicted real value between 0 and 1, to be converted to a binary equivalent of 0 or 1 either by rounding or a cut-off value.

The LogisticRegression class of the sklean.linear_model module implements logistic regression. Let's implement this classifier model by writing a new class named LogisticRegressionModel that extends LinearRegressionModel with the following code:

```
In [ ]:
    from sklearn.linear_model import LogisticRegression

    class LogisticRegressionModel(LinearRegressionModel):
        def get_model(self):
            return LogisticRegression(solver='lbfgs')
```

The same underlying linear regression logic is used in our new classifier model. The get_model() method is overridden to return an instance of the LogisticRegression classifier model, using the LBFGS solver algorithm in the optimization problem.

 A paper on the **limited-memory Broyden–Fletcher–Goldfarb–Shanno (LBFGS)** algorithm for machine learning can be read at https://arxiv.org/pdf/1802.05374.pdf

Create an instance of this model and provide our data:

```
In [ ]:
    logistic_reg_model = LogisticRegressionModel()
    logistic_reg_model.learn(df_input, y_direction, start_date='2018',
                            end_date='2019', lookback_period=100)
```

Again, the parameter values indicate that we are interested in performing predictions for the year of 2018, and we will be using a `lookback_period` value of `100` as the number of daily historical data points when fitting our model. Let's inspect the results stored in `df_result` with the `head()` command:

```
In [ ]:
    logistic_reg_model.df_result.head()
```

This produces the following table:

Date	Actual	Predicted
2018-01-02	True	True
2018-01-03	True	True
2018-01-04	True	True
2018-01-05	False	True
2018-01-08	True	True

Since our target variables are Boolean values, the model outputs predict Boolean values as well. But how well does our model perform? In the following sections, we will explore risk metrics for measuring our predictions. These metrics are different from those used for regression-based predictions in earlier sections. Classification-based machine learning takes another approach for measuring output labels.

Risk metrics for measuring classification-based predictions

In this section, we will explore common risk metrics for measuring classification-based machine learning predictions, namely the confusion matrix, accuracy score, precision score, recall score, and F1 score.

Confusion matrix

A confusion matrix, or error matrix, is a square matrix that helps to visualize and describe the performance of a classification model for which the true values are known. The `confusion_matrix` function of the `sklearn.metrics` module helps to calculate this matrix for us, as shown in the following code:

```
In [ ]:
    from sklearn.metrics import confusion_matrix

    df_result = logistic_reg_model.df_result
    actual = list(df_result['Actual'])
    predicted = list(df_result['Predicted'])

    matrix = confusion_matrix(actual, predicted)
In [ ]:
    print(matrix)
Out[ ]:
    [[60 66]
     [55 70]]
```

We obtain the actual and predicted values as separate lists. Since we have two types of class labels, we obtain a two-by-two matrix. The `heatmap` module of the `seaborn` library helps us understand this matrix.

> Seaborn is a data visualization library based on Matplotlib. It provides a high-level interface for drawing attractive and informative statistical graphics, and is a popular tool for data scientists. If you do not have Seaborn installed, simply run the command: `pip install seaborn`

Run the following Python codes to generate the confusion matrix:

```
In [ ]:
    %matplotlib inline
    import seaborn as sns
    import matplotlib.pyplot as plt

    plt.subplots(figsize=(12,8))
    sns.heatmap(matrix.T, square=True, annot=True, fmt='d', cbar=False,
                xticklabels=flags, yticklabels=flags)
    plt.xlabel('Actual')
    plt.ylabel('Predicted')
    plt.title('JPM percentage returns 2018');
```

This produces the following output:

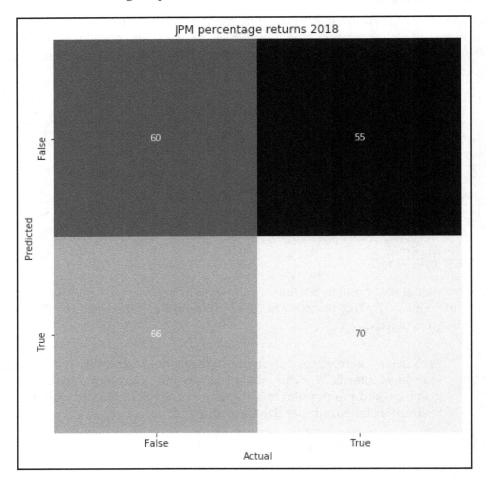

Don't let the confusion matrix confuse you. Let's break down the numbers in a logical manner and see how easily a confusion matrix works. Starting from the left column, we have a total of 126 samples classified as **False**, of which the classifier predicted correctly **60** times, and these are known as **true negatives** (**TNs**). However, the classifier predicted it wrongly **66** times, and these are known as **false negatives** (**FNs**). In the right column, we have a total of 125 samples belonging to the **True** class. The classifier predicted wrongly **55** times, and these are known as **false positives** (**FPs**). The classifier did predict correctly **70** times though, and these are known as **true positives** (**TPs**). These computed rates are used in other risk metrics, as we shall discover in the following sections.

Accuracy score

An accuracy score is the ratio of correct predictions to the total number of observations. By default, it is expressed as a fractional value between 0 and 1. When the accuracy score is 1.0, it means that the entire set of predicted labels in the sample matches with the true set of labels. The accuracy score can be written as follows:

$$accuracy(y, \hat{y}) = \frac{1}{n} \sum_{i=0}^{n-1} I(\hat{y}_i = y_i)$$

Here, $I(x)$ is the indicator function that returns 1 for a correct prediction, and 0 otherwise. The `accuracy_score` function of the `sklearn.metrics` module calculates this score for us with the following code:

```
In [ ]:
    from sklearn.metrics import accuracy_score
    print('accuracy_score:', accuracy_score(actual, predicted))
Out[ ]:
    accuracy_score: 0.5179282868525896
```

The accuracy score suggests that our model is correct 52% of the time. Accuracy scores are great at measuring symmetrical datasets where values of false positives and false negatives are almost the same. To evaluate the performance of our model fully, we need to look at other risk metrics.

Precision score

A precision score is the ratio of correctly predicted positive observations to the total number of predicted positive observations, and can be written as follows:

$$\text{Precision score} = \frac{\text{true positives}}{\text{true positives} + \text{false positives}}$$

This gives a precision score between 0 and 1, with 1 as the best value indicating that the model classifies correctly all the time. The `precision_score` function of the `sklearn.metrics` module calculates this score for us with the following code:

```
In [ ]:
    from sklearn.metrics import precision_score
    print('precision_score:', precision_score(actual, predicted))
Out[ ]:
    precision_score: 0.5147058823529411
```

The precision score suggests that our model is able to predict a classification correctly 52% of the time.

Recall score

The recall score is the ratio of correctly predicted positive observations to all the observations in the actual class, and can be written as follows:

$$\text{Recall score} = \frac{\text{true positives}}{\text{true positives} + \text{false negatives}}$$

This gives a recall score of between 0 and 1, with 1 as the best value. The `recall_score` function of the `sklearn.metrics` module calculates this score for us with the following code:

```
In [ ]:
    from sklearn.metrics import recall_score
    print('recall_score:', recall_score(actual, predicted))
Out[ ]:
    recall_score: 0.56
```

The recall score suggests that our logistic regression model correctly identifies positive samples 56% of the time.

F1 score

The F1 score, or F-measure, is the weighted average of the precision score and the recall score, and can be written as follows:

$$F1 = 2 * \frac{precision * recall}{precision + recall}$$

This gives an F1 score between 0 and 1. When either the precision score or the recall score is 0, the F1 score will be 0. However, when both the precision score and recall score are positive, the F1 score gives equal weights to both measures. Maximizing the F1 score creates a balanced classification model with optimal balance of recall and precision.

The `f1_score` function of the `sklearn.metrics` module calculates this score for us with the following code:

```
In [ ]:
    from sklearn.metrics import f1_score
    print('f1_score:', f1_score(actual, predicted))
Out[ ]:
    f1_score: 0.5363984674329502
```

The F1 score of our logistic regression model is 0.536.

Support vector classifier

A **support vector classifier** (**SVC**) is a concept of a **support vector machine** (**SVM**) that uses support vectors for classifying datasets.

 More information on SVMs can be found at `http://www.statsoft.com/ textbook/support-vector-machines`.

The `SVC` class of the `sklean.svm` module implements the SVM classifier. Write a class named `SVCModel` and extend `LogisticRegressionModel` with the following code:

```
In [ ]:
    from sklearn.svm import SVC

    class SVCModel(LogisticRegressionModel):
        def get_model(self):
            return SVC(C=1000, gamma='auto')
In [ ]:
    svc_model = SVCModel()
    svc_model.learn(df_input, y_direction, start_date='2018',
                    end_date='2019', lookback_period=100)
```

Here, we are overriding the `get_model()` method to return the `SVC` class of scikit-learn. A high-penalty `C` value of `1000` is specified. The `gamma` parameter is the kernel coefficient with a default value of `auto`. The `learn()` command is executed with our usual model parameters. With that, let's run the risk metrics on this model:

```
In [ ]:
    df_result = svc_model.df_result
    actual = list(df_result['Actual'])
    predicted = list(df_result['Predicted'])
In [ ]:
```

```
        print('accuracy_score:', accuracy_score(actual, predicted))
        print('precision_score:', precision_score(actual, predicted))
        print('recall_score:', recall_score(actual, predicted))
        print('f1_score:', f1_score(actual, predicted))
Out[ ]:
    accuracy_score: 0.5577689243027888
    precision_score: 0.5538461538461539
    recall_score: 0.576
    f1_score: 0.5647058823529412
```

We obtain better scores than from the logistic regression classifier model. By default, the C value of the linear SVM is 1.0, which would in practice give us generally comparable performance with the logistic regression model. There is absolutely no rule of thumb for choosing a C value, as it depends entirely on the training dataset. A nonlinear SVM kernel can be considered by supplying a `kernel` parameter to the `SVC()` model. More information on SVM kernels is available at `https://scikit-learn.org/stable/modules/svm.html#svm-kernels`.

Other types of classifiers

Besides logistic regression and SVC, scikit-learn contains many other types of classifiers for machine learning. The following sections discuss some classifiers that we can also consider implementing in our classification-based model.

Stochastic gradient descent

Stochastic gradient descent (SGD) is a form of **gradient descent** that works by using an iterative process to estimate the gradient towards minimizing an objective loss function, such as a linear support vector machine or logistic regression. The stochastic term comes about as samples are chosen at random. When lesser iterations are used, bigger steps are taken to reach the solution, and the model is said to have a **high learning rate**. Likewise, with more iterations, smaller steps are taken, resulting in a model with a **small learning rate**. SGD is a popular choice of machine learning algorithm among practitioners as it has been effectively used in large-scale text classification and natural language processing models.

The `SGDClassifier` class of the `sklearn.linear_model` module implements the SGD classifier.

Linear discriminant analysis

Linear discriminant analysis (**LDA**) is a classic classifier that uses a linear decision surface, where the mean and variance for every class of the data is estimated. It assumes that the data is Gaussian, and that each attribute has the same variance, and values of each variable are around the mean. LDA computes *discriminant scores* by using Bayes' theorem for each observation to determine to which class it belongs.

The `LinearDiscriminantAnalysis` class of the `sklearn.discriminant_analysis` module implements the LDA classifier.

Quadratic discriminant analysis

Quadratic discriminant analysis (**QDA**) is very similar to LDA, but uses a quadratic decision boundary and each class uses its own estimate of variance. Running the risk metrics shows that the QDA model does not necessarily give better performance than the LDA model. The type of decision boundary has to be taken into consideration for the model required. QDA is better suited for large datasets, as it tends to have a lower bias and higher variance. On the other hand, LDA is suitable for smaller datasets that have a lower bias and a higher variance.

The `QuadraticDiscriminantAnalysis` class of the `sklearn.discriminant_analysis` module implements the QDA model.

KNN classifier

The **k-nearest neighbors** (**k-NN**) classifier is a simple algorithm that conducts a simple majority vote of the nearest neighbors of each point, and that point is assigned to a class that has the most representatives within the nearest neighbors of the point. While there is not a need to train a model for generalization, the predicting phase is slower and costlier in terms of time and memory.

The `KNeighborsClassifier` class of the `sklearn.neighbors` module implements the KNN classifier.

Conclusion on the use of machine learning algorithms

You may have observed that predicted values from our models are far off from actual values. This chapter aims to demonstrate the best of the machine learning features that scikit-learn offers, which may possibly be used to predict time series data. No studies to date have shown that machine learning algorithms can predict prices even close to 100% of the time. A lot more effort goes into building and running machine learning systems effectively.

Summary

In this chapter, we have been introduced to machine learning in the context of finance. We discussed how AI and machine learning is transforming the financial sector. Machine learning can be supervised or unsupervised, and supervised algorithms can be regression-based and classification-based. The scikit-learn Python library provides various machine learning algorithms and risk metrics.

We discussed the use of regression-based machine learning models such as OLS regression, ridge regression, LASSO regression, and elastic net regularization in predicting continuous values such as security prices. An ensemble of decision trees was also discussed, such as the bagging regressor, gradient tree boosting, and random forests. To measure the performance of regression models, we visited the MSE, MAE, explained variance score, and R^2 score.

Classification-based machine learning classifies input values as classes or labels. Such classes may be bi-class or multi-class. We discussed the use of logistic regression, SVC, LDA and QDA, and k-NN classifiers for predicting price trends. To measure the performance of classification models, we visited the confusion matrix, accuracy score, precision and recall scores, as well as the F1 score.

In the next chapter, we will explore the use of deep learning in finance.

11
Deep Learning for Finance

Deep learning represents the very cutting edge of **Artificial Intelligence (AI)**. Unlike machine learning, deep learning takes a different approach in making predictions by using a neural network. An artificial neural network is modeled on the human nervous system, consisting of an input layer and an output layer, with one or more hidden layers in between. Each layer consists of artificial neurons working in parallel and passing outputs to the next layer as inputs. The word *deep* in deep learning comes from the notion that as data passes through more hidden layers in an artificial neural network, more complex features can be extracted.

TensorFlow is an open source, powerful machine learning and deep learning framework developed by Google. In this chapter, we will take a hands-on approach to learning TensorFlow by building a deep learning model with four hidden layers to predict the prices of a security. Deep learning models are trained by passing the entire dataset forward and backward through the network, with each iteration known as an **epoch**. Because the input data can be too big to be fed, training can be done in batches, and this process is known as **mini-batch training**.

Another popular deep learning library is Keras, which utilizes TensorFlow as the backend. We will also take a hands-on approach to learning Keras and see how easy it is to build a deep learning model to predict credit card payment defaults.

In this chapter, we will cover the following topics:

- An introduction to neural networks
- Neurons, activation functions, loss functions, and optimizers
- Different types of neural network architectures
- How to build security price prediction deep learning model using TensorFlow
- Keras, a user-friendly deep learning framework
- How to build credit card payment default prediction deep learning model using Keras
- How to display recorded events in a Keras history

A brief introduction to deep learning

The theory behind deep learning began as early as the 1940s. However, its popularity has soared in recent years thanks in part to improvements in computing hardware technology, smarter algorithms, and the adoption of deep learning frameworks. There is much to cover beyond this book. This section serves as a quick guide to gain a working knowledge for following the examples that we will cover in later parts of this chapter.

What is deep learning ?

In `Chapter 10`, *Machine Learning for Finance*, we learned how machine learning is useful for making predictions. Supervised learning uses error-minimization techniques to fit a model with training data, and can be regression based or classification based.

Deep learning takes a different approach in making predictions by using a neural network. Modeled on the human brain and the nervous system, an artificial neural network consists of a hierarchy of layers, with each layer made up of many simple units known as neurons, working in parallel and transforming the input data into abstract representations as the output data, which are fed to the next layer as input. The following diagram illustrates an artificial neural network:

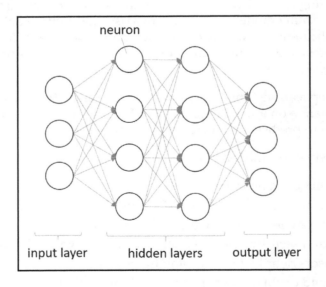

Artificial neural networks consist of three types of layers. The first layer that accepts input is known as the **input layer**. The last layer where output is collected is known as the **output layer**. The layers between the input and output layers are known as **hidden layers**, since they are hidden from the interface of the network. There can be many combinations of hidden layers performing different activation functions. Naturally, more complex computations lead to a rise in demand for more powerful machines, such as the GPUs required to compute them.

The artificial neuron

An artificial neuron receives one or more input and are multiplied by values known as **weights**, summed up and passed to an activation function. The final values computed by the activation function makes up the neuron's output. A bias value may be included in the summation term to help fit the data. The following diagram illustrates an artificial neuron:

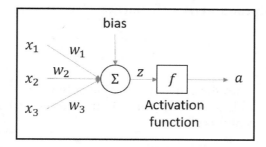

The summation term can be written as a linear equation such that $Z = x_1 w_1 + x_2 w_2 + ... + b$. The neuron uses a nonlinear activation function f to transform the input to become the output a, and can be written as $a = f(Z)$.

Activation function

An activation function is part of an artificial neuron that transforms the sum of weighted inputs into another value for the next layer. Usually, the range of this output value is -1 or 0 to 1. An artificial neuron is said to be activated when it passes a non-zero value to another neuron. There are several types of activation functions, mainly:

- Linear
- Sigmoid
- Tanh
- Hard tanh
- Rectified linear unit
- Leaky ReLU
- Softplus

For example, a **rectified linear unit** (**ReLU**) function is written as:

$$f(x) = \begin{cases} 0 & \text{if } x \leq 0 \\ x & \text{otherwise} \end{cases}$$

The ReLU activates a node with the same input value only when the input is above zero. Researchers prefer to use ReLU as it trains better than sigmoid activation functions. We will be using ReLU in later parts of this chapter.

In another example, the leaky ReLU is written as:

$$f(x) = \begin{cases} 0.01x & \text{if } x \leq 0 \\ x & \text{otherwise} \end{cases}$$

The leaky ReLU addresses the issue of a dead ReLU when $x \leq 0$ by having a small negative slope around 0.01 when x is zero and below.

Loss functions

The loss function computes the error between the predicted value of a model and the actual value. The smaller the error value, the better the model is in prediction. Some loss functions used in regression-based models are:

- **Mean squared error (MSE)** loss
- **Mean absolute error (MAE)** loss
- Huber loss
- Quantile loss

Some loss functions used in classification-based models are:

- Focal loss
- Hinge loss
- Logistic loss
- Exponential loss

Optimizers

Optimizers help to tweak the model weights optimally in minimizing the loss function. There are several types of optimizers that you may come across in deep learning:

- **AdaGrad (adaptive gradient)**
- **Adam (adaptive moment estimation)**
- **LBFGS (limited-memory Broyden-Fletcher-Goldfarb-Shannon)**
- **Rprop (resilient backpropagation)**
- **RMSprop (root mean square propagation)**
- **SGD (stochastic gradient descent)**

Adam is a popular choice of optimizer, and is seen as a combination of RMSprop and SGD with momentum. It is an adaptive learning rate optimization algorithm, computing individual learning rates for different parameters.

Network architecture

The network architecture of a neural network defines its behavior. There are many forms of network architecture available; some are:

- **Perceptron (P)**
- **Feed forward (FF)**
- **Deep feed forward (DFF)**
- **Radial basis function network (RBF)**
- **Recurrent neural network (RNN)**
- **Long/short-term memory (LSTM)**
- **Autoencoder (AE)**
- **Hopfield network (HN)**
- **Boltzmann machine (BM)**
- **Generative adversarial network (GAN)**

The most well-known and easy-to-understand neural network is the feed forward multilayer neural network. It can represent any function using an input layer, one or more hidden layers, and a single output layer. A list of neural networks can be found at http://www.asimovinstitute.org/neural-network-zoo/.

TensorFlow and other deep learning frameworks

TensorFlow is a free and open source library from Google, available in Python, C++, Java, Rust, and Go. It contains various neural networks for training deep learning models. TensorFlow can be applied to various scenarios, such as image classification, malware detection, and speech recognition. The official page for TensorFlow is https://www.tensorflow.org.

Other popular deep learning frameworks used in the industry are Theano, PyTorch, CNTK (Microsoft Cognitive Toolkit), Apache MXNet, and Keras.

What is a tensor ?

The *Tensor* in TensorFlow indicates that the frameworks define and run computations involving tensors. A tensor is nothing more than a type of n-dimensional vector with certain transformative properties. A non-dimensional tensor is a scalar or number. A one-dimensional tensor is a vector. A two-dimensional tensor is a matrix. Tensors offer more natural representations of data, for example in images in the field of computer vision.

The basic properties of vector spaces and the elementary mathematical properties of tensors make them particularly useful in physics and engineering.

A deep learning price prediction model with TensorFlow

In this section, we will learn how to use TensorFlow as a deep learning framework in building a price prediction model. Five years of pricing data, from 2013 to 2017, will be used for training our deep learning model. We will attempt to predict the prices of Apple (AAPL) in the following year of 2018.

Feature engineering our model

The daily adjusted closing prices of our data make up the target variables. The independent variables defining the features of our model are made up of these technical indicators:

- **Relative strength index (RSI)**
- **Williams %R (WR)**
- **Awesome oscillator (AO)**
- **Volume-weighted average price (VWAP)**
- **Average daily trading volume (ADTV)**
- 5-day **moving average (MA)**
- 15-day moving average
- 30-day moving average

This gives us eight features for our model.

Requirements

You should have NumPy, pandas, Jupyter, and scikit-learn libraries installed, as mentioned in previous chapters. The following sections highlight additional important requirements for building our deep learning model.

Intrinio as our data provider

Intrinio (`https://intrinio.com/`) is a premium API-based financial data provider. We will be using the US Fundamentals and Stock Prices subscription, which gives us access to US historical stock prices and well-calculated technical indicator values. After registering for an account, your API keys can be found in your account settings, which we will use later.

Compatible Python environment for TensorFlow

At the time of writing, the latest stable version of TensorFlow is r1.13. This version is compatible with Python 2.7, 3.4, 3.5, and 3.6. As the preceding chapters in this book use Python 3.7, we need to set up a separate Python 3.6 environment for running the examples in this chapter. The virtualenv tool (`https://virtualenv.pypa.io/`) is recommended to isolate Python environments.

The requests library

The `requests` Python library is required to help us make HTTP calls to Intrinio APIs. The official web page for `requests` is `http://docs.python-requests.org/en/master/`. Install `requests` by running this command in your terminal: `pip install requests`.

The TensorFlow library

There are a number of variants of TensorFlow available for installation. You may choose between CPU-only or GPU support versions, alpha versions, and nightly versions. More installation instructions are available at `https://www.tensorflow.org/install/pip`. At a minimum, the following terminal command installs the latest CPU-only stable version of TensorFlow: `pip install tensorflow`.

Downloading the dataset

This section describes the steps for downloading our required prices and technical indicator values from Intrinio. Comprehensive documentation on the API calls can be found at `https://docs.intrinio.com/documentation/api_v2`. If you decide to use another data provider, go ahead and skip this section:

1. Write a `query_intrinio()` function that will make an API call to Intrinio, with the following codes:

   ```
   In [ ]:
       import requests

       BASE_URL = 'https://api-v2.intrinio.com'

       # REPLACE YOUR INTRINIO API KEY HERE!
       INTRINIO_API_KEY =
   'Ojc3NjkzOGNmNDMxMGFiZWZiMmMxMmY0Yjk3MTQzYjdh'

       def query_intrinio(path, **kwargs):
           url = '%s%s'%(BASE_URL, path)
           kwargs['api_key'] = INTRINIO_API_KEY
           response = requests.get(url, params=kwargs)

           status_code = response.status_code
           if status_code == 401:
               raise Exception('API key is invalid!')
           if status_code == 429:
               raise Exception('Page limit hit! Try again in 1
   minute')
           if status_code != 200:
               raise Exception('Request failed with status
   %s'%status_code)

           return response.json()
   ```

   This function accepts the `path` and `kwargs` parameters. The `path` parameter refers to the specific Intrinio API context path. The `kwargs` keyword argument is a dictionary that gets passed along to a HTTP GET request call as request parameters. The API key is inserted into this dictionary on every API call to identify the user account. Any API responses are expected to be in JSON format with a HTTP status code of 200; otherwise, an exception will be thrown.

2. Write a `get_technicals()` function to download technical indicator values from Intrinio, with the following codes:

```
In [ ]:
    import pandas as pd
    from pandas.io.json import json_normalize

    def get_technicals(ticker, indicator, **kwargs):
        url_pattern = '/securities/%s/prices/technicals/%s'
        path = url_pattern%(ticker, indicator)
        json_data = query_intrinio(path, **kwargs)

        df = json_normalize(json_data.get('technicals'))
        df['date_time'] = pd.to_datetime(df['date_time'])
        df = df.set_index('date_time')
        df.index = df.index.rename('date')
        return df
```

The `ticker` and `indicator` parameters make up the API context path for downloading the specific indicator of a security. The response is expected to be in JSON format, with a key named `technicals` containing the list of technical indicator values. The `json_normalize()` function of pandas helps to convert these values into a flat table DataFrame object. Extra formatting is required to set date and time values as the index under the `date` name.

3. Define the values for the request parameters:

```
In [ ]:
    ticker = 'AAPL'
    query_params = {'start_date': '2013-01-01', 'page_size': 365*6}
```

We shall be querying data for the security `AAPL`, from 2013 to 2018, inclusive. The big `page_size` value gives us sufficient space to request six years of data in a single query.

4. Run the following commands at one-minute intervals each to download the technical indicator data:

```
In [ ]:
    df_rsi = get_technicals(ticker, 'rsi', **query_params)
    df_wr = get_technicals(ticker, 'wr', **query_params)
    df_vwap = get_technicals(ticker, 'vwap', **query_params)
    df_adtv = get_technicals(ticker, 'adtv', **query_params)
    df_ao = get_technicals(ticker, 'ao', **query_params)
    df_sma_5d = get_technicals(ticker, 'sma', period=5,
**query_params)
```

```
    df_sma_5d = df_sma_5d.rename(columns={'sma':'sma_5d'})
    df_sma_15d = get_technicals(ticker, 'sma', period=15,
**query_params)
    df_sma_15d = df_sma_15d.rename(columns={'sma':'sma_15d'})
    df_sma_30d = get_technicals(ticker, 'sma', period=30,
**query_params)
    df_sma_30d = df_sma_30d.rename(columns={'sma':'sma_30d'})
```

Beware of paging limits when performing Intrinio API queries! API requests with a `page_size` greater than 100 are subjected to a per-minute request limit. If a call fails with status code 429, try again in one minute. Information on Intrinio's limits can be found at `https://docs.intrinio.com/documentation/api_v2/limits`.

This gives us eight variables, each containing the DataFrame object of the respective technical indicator values. The MA data columns are renamed to avoid a naming conflict when joining the data later.

5. Write a `get_prices()` function to download the historical prices of a security:

```
In [ ]:
    def get_prices(ticker, tag, **params):
        url_pattern = '/securities/%s/historical_data/%s'
        path = url_pattern%(ticker, tag)
        json_data = query_intrinio(path, **params)

        df = json_normalize(json_data.get('historical_data'))
        df['date'] = pd.to_datetime(df['date'])
        df = df.set_index('date')
        df.index = df.index.rename('date')
        return df.rename(columns={'value':tag})
```

The `tag` parameter specifies the data tag of the security to download. The JSON response is expected to contain a key named `historical_data` containing the list of values. The column containing the prices in the DataFrame object is renamed from `value` to its data tag.

The Intrinio data tags are used to download specific values from the system. The list of data tags available with explanations can be found at `https://data.intrinio.com/data-tags/all`.

6. Using the `get_prices()` function, download the adjusted closing prices of AAPL:

```
In [ ]:
    df_close = get_prices(ticker, 'adj_close_price',
**query_params)
```

7. As the features are used to predict the next day's closing prices, we need to shift the prices backwards by one day to align this mapping. Create the target variables:

```
In [ ]:
    df_target = df_close.shift(1).dropna()
```

8. Finally, combine all the DataFrame objects together with the `join()` command and drop the empty values:

```
In [ ]:
    df = df_rsi.join(df_wr).join(df_vwap).join(df_adtv)\
        .join(df_ao).join(df_sma_5d).join(df_sma_15d)\
        .join(df_sma_30d).join(df_target).dropna()
```

Our dataset is now ready, contained in the `df` DataFrame. We can proceed to split the data for training.

Scaling and splitting the data

We are interested in using the earliest five years of pricing data for training our model, and the most recent year of 2018 for testing our predictions. Run the following codes to split our `df` dataset:

```
In [ ]:
    df_train = df['2017':'2013']
    df_test = df['2018']
```

The `df_train` and `df_test` variables contain our training and testing data respectively.

An important step in data preprocessing is to normalize the dataset. This will transform input feature values to a mean of zero and a variance of one. Normalization helps to avoid biases during training due to the different scales of input features.

The `MinMaxScaler` function of the `sklearn` module helps to transform each feature into a range between -1 and 0, with the following codes:

```
In [ ]:
    from sklearn.preprocessing import MinMaxScaler

    scaler = MinMaxScaler(feature_range=(-1, 1))
    train_data = scaler.fit_transform(df_train.values)
    test_data = scaler.transform(df_test.values)
```

The `fit_transform()` function computes the parameters for scaling and transforms the data, while the `transform()` function only transforms the data by reusing the computed parameters.

Next, split the scaled training dataset into independent and target variables. The target values are on the last column, with the remaining columns as the features:

```
In [ ]:
    x_train = train_data[:, :-1]
    y_train = train_data[:, -1]
```

Do the same on our testing data for features only:

```
In [ ]:
    x_test = test_data[:, :-1]
```

With our training and testing dataset prepared, let's begin to build an artificial neural network with TensorFlow.

Building an artificial neural network with TensorFlow

This section walks you through the process of setting up an artificial neural network for deep learning with four hidden layers. There are two phases involved; first in assembling the graph, and next in training the model.

Phase 1 – assembling the graph

The following steps describe the process of setting up a TensorFlow graph:

1. Create placeholders for inputs and labels with the following codes:

```
In [ ]:
    import tensorflow as tf

    num_features = x_train.shape[1]

    x = tf.placeholder(dtype=tf.float32, shape=[None,
num_features])
    y = tf.placeholder(dtype=tf.float32, shape=[None])
```

A TensorFlow operation starts with placeholders. Here, we defined two placeholders x and y for containing the network inputs and outputs, respectively. The shape parameter defines the shape of the tensor to be fed, with None meaning that the number of observations is unknown at this point. The second dimension of x is the number of features that we have, reflected in the num_features variable. Later, as we shall see, placeholder values are fed using the feed_dict command.

2. Create weight and bias initializers for hidden layers. Our model will consist of four hidden layers. The first layer contains 512 neurons, about three times the size of the input. The second, third, and fourth layers contain 256, 128, and 64 neurons, respectively. The reduction of the number of neurons in subsequent layers compresses the information in the network.

Initializers are used to initialize the network variables before training. It is important to use proper initialization at the start of the optimization problem to produce good solutions to the underlying problem. The use of a variance scaling initializer and a zeros initializer is demonstrated with the following code:

```
In [ ]:
    nl_1, nl_2, nl_3, nl_4 = 512, 256, 128, 64

    wi = tf.contrib.layers.variance_scaling_initializer(
        mode='FAN_AVG', uniform=True, factor=1)
    zi = tf.zeros_initializer()

    # 4 Hidden layers
    wt_hidden_1 = tf.Variable(wi([num_features, nl_1]))
    bias_hidden_1 = tf.Variable(zi([nl_1]))
```

```
wt_hidden_2 = tf.Variable(wi([nl_1, nl_2]))
bias_hidden_2 = tf.Variable(zi([nl_2]))

wt_hidden_3 = tf.Variable(wi([nl_2, nl_3]))
bias_hidden_3 = tf.Variable(zi([nl_3]))

wt_hidden_4 = tf.Variable(wi([nl_3, nl_4]))
bias_hidden_4 = tf.Variable(zi([nl_4]))

# Output layer
wt_out = tf.Variable(wi([nl_4, 1]))
bias_out = tf.Variable(zi([1]))
```

Besides placeholders, variables within TensorFlow are updated during graph execution. Here, the variables are the weights and bias that will change during training. The `variance_scaling_initializer()` command returns an initializer that generates tensors for our weights without scaling variance. The `FAN_AVG` mode indicates to the initializer to use the average number of input and output connections, with the `uniform` parameter as `True` to use uniform random initialization and a scale factor of 1. This is akin to training DFF neural networks.

In **multilayer perceptrons (MLP)** such as our model, the first dimension of the weights layer is the same as the second dimension of the previous weights layer. The bias dimensions correspond to the number of neurons in the current layer. The neuron of the last layer is expected to have only one output.

3. Now is the time to combine our placeholder inputs with weights and bias for the four hidden layers using the following code:

```
In [ ]:
    hidden_1 = tf.nn.relu(
        tf.add(tf.matmul(x, wt_hidden_1), bias_hidden_1))
    hidden_2 = tf.nn.relu(
        tf.add(tf.matmul(hidden_1, wt_hidden_2), bias_hidden_2))
    hidden_3 = tf.nn.relu(
        tf.add(tf.matmul(hidden_2, wt_hidden_3), bias_hidden_3))
    hidden_4 = tf.nn.relu(
        tf.add(tf.matmul(hidden_3, wt_hidden_4), bias_hidden_4))
    out = tf.transpose(tf.add(tf.matmul(hidden_4, wt_out),
bias_out))
```

The tf.matmul command multiplies the input and weight matrices, adding the bias values using the tf.add command. Each hidden layer of the neural network is transformed by an activation function. In this model, we are using ReLU as the activation function for all layers using the tf.nn.relu command. The output of each hidden layer is fed to the input of the next hidden layer. The last layer, which is the output layer with a single vector output, must be transposed with the tf.transpose command.

4. Specify the loss function of the network to measure the error between predicted and actual values during training. For regression-based models such as ours, the MSE is commonly used:

```
In [ ]:
    mse = tf.reduce_mean(tf.squared_difference(out, y))
```

The tf.squared_difference command is defined to return the squared errors between the predicted and actual values, and the tf.reduce_mean command is the loss function for minimizing the mean during training.

5. Create the optimizer with the following code:

```
In [ ]:
    optimizer = tf.train.AdamOptimizer().minimize(mse)
```

In minimizing the loss function, an optimizer helps to compute the network weight and bias during training. Here, we are using the Adam algorithm with default values. With this important step completed, we may now embark on phase two in training our model.

Phase 2 – training our model

The following steps describe the process of training our model:

1. Create a TensorFlow Session object to encapsulate the environment in which a neural network model operates:

```
In [ ]:
    session = tf.InteractiveSession()
```

Here, we are specifying a session for use in an interactive context, in this case a Jupyter notebook. A regular `tf.Session` is non-interactive and requires an explicit `Session` object to be passed using the `with` keyword when running operations. `InteractiveSession` removes this need and is more convenient as it reuses the `session` variable.

2. TensorFlow requires that all global variables are to be initialized before training. Do this using the `session.run` command:

```
In [ ]:
    session.run(tf.global_variables_initializer())
```

3. Run the following codes to train our model using mini-batch training:

```
In [ ]:
    from numpy import arange
    from numpy.random import permutation

    BATCH_SIZE = 100
    EPOCHS = 100

    for epoch in range(EPOCHS):
        # Shuffle the training data
        shuffle_data = permutation(arange(len(y_train)))
        x_train = x_train[shuffle_data]
        y_train = y_train[shuffle_data]

        # Mini-batch training
        for i in range(len(y_train)//BATCH_SIZE):
            start = i*BATCH_SIZE
            batch_x = x_train[start:start+BATCH_SIZE]
            batch_y = y_train[start:start+BATCH_SIZE]
            session.run(optimizer, feed_dict={x: batch_x, y:
batch_y})
```

An epoch is a single iteration of the entire dataset being passed forward and backward through the network. Usually several epochs are performed on different permutations of the training data for the network to learn its behavior. There is no fixed number of epochs for a good model, as it depends on how diverse the data is. Because the dataset can be too big to be fed into the model in one epoch, mini-batch training divides the dataset into parts and feeds it into the `session.run` command for learning. The first parameter specifies the optimization algorithm instance. The `feed_dict` parameter is given a dictionary containing our x and y placeholders mapped to batches of our independent and target values respectively.

4. After our model is fully trained, use it for prediction with our testing data containing the features:

```
In [ ]:
    [predicted_values] = session.run(out, feed_dict={x: x_test})
```

The `session.run` command is called with the first parameter as the output layer transformation function. The `feed_dict` parameter is fed with our testing data. The first item in the output list is read as the final output predicted values.

5. As the predicted values are also normalized, we need to scale them back to the original values:

```
In [ ]:
    predicted_scaled_data = test_data.copy()
    predicted_scaled_data[:, -1] = predicted_values
    predicted_values =
scaler.inverse_transform(predicted_scaled_data)
```

Create a copy of our initial training data with the `copy()` command onto the new `predicted_scaled_data` variable. The last column will be replaced with our predicted values. Next, the `inverse_transform()` command scales our data back to the original size, giving us the predicted values for comparison with actual observed values.

Plotting predicted and actual values

Let's plot the predicted and actual values onto a graph to visualize the performance of our deep learning model. Run the following codes to extract our values of interest:

```
In [ ]:
    predictions = predicted_values[:, -1][::-1]
    actual = df_close['2018']['adj_close_price'].values[::-1]
```

The rescaled `predicted_values` dataset is a NumPy `ndarray` object with predicted values on the last column. These values and the actual adjusted closing prices of 2018 are extracted to the `predictions` and `actual` variables respectively. Since the format of the original dataset is in descending order of time, we reverse them in ascending order for plotting on a graph. Run the following codes to generate a graph:

```
In [ ]:
    %matplotlib inline
    import matplotlib.pyplot as plt
```

```
plt.figure(figsize=(12,8))
plt.title('Actual and predicted prices of AAPL 2018')
plt.plot(actual, label='Actual')
plt.plot(predictions, linestyle='dotted', label='Predicted')
plt.legend()
```

The following output is produced:

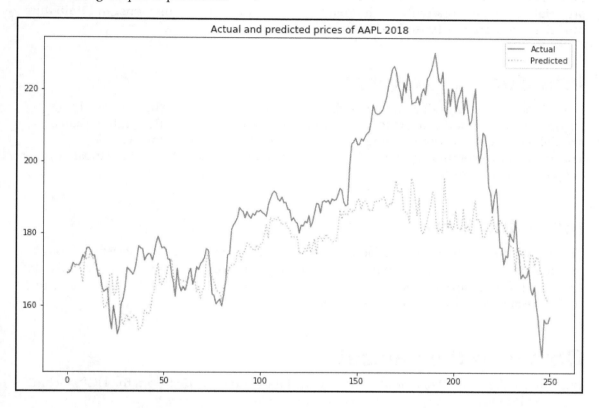

The solid line shows the actual adjusted closing prices, while the dotted lines show the predicted prices. Notice how our predictions follow a general trend with actual prices even though the model did not have any knowledge of the actual prices in 2018. Still, there is plenty of room for improvements in our deep learning prediction model, such as in the design of the neuron network architecture, hidden layers, activation functions, and initialization schemes.

Credit card payment default prediction with Keras

Another popular deep learning Python library is Keras. In this section, we will use Keras to build a credit card payment default prediction model, and see how easy it is to construct an artificial neural network with five hidden layers, apply activation functions, and train this model as compared to TensorFlow.

Introduction to Keras

Keras is an open source deep learning library in Python, designed to be high level, user friendly, modular, and extensible. Keras was conceived to be an interface rather than a standalone machine learning framework, running on top of TensorFlow, CNTK, and Theano. Its huge community base with over 200,000 users makes it one of the most popular deep learning libraries.

Installing Keras

The official documentation page for Keras is at `https://keras.io`. The easiest way to install Keras is running this command in your terminal: `pip install keras`. By default, Keras will use TensorFlow as its tensor manipulation library, though it is also possible to configure another backend implementation.

Obtaining the dataset

We will use the Default of Credit Card Clients dataset downloaded from the UCI Machine Learning Repository (`https://archive.ics.uci.edu/ml/datasets/default+of+credit+card+clients`). Source: Yeh, I. C., and Lien, C. H. (2009). *The comparisons of data mining techniques for the predictive accuracy of probability of default of credit card clients. Expert Systems with Applications, 36(2), 2473-2480.*

This dataset contains customer default payments in Taiwan. Refer to the section Attribute Information on the web page for the naming conventions used for the columns in the dataset. As the original dataset is in Microsoft Excel Spreadsheet XLS format, additional data processing is required. Open the file and remove the first row and first column containing supplementary attribute information, and save it as a CSV file. A copy of this file is found in `files\chapter11\default_cc_clients.csv` of the source code repository.

Read this dataset as a `pandas` DataFrame object to a new variable named `df`:

```
In [ ]:
    import pandas as pd

    df = pd.read_csv('files/chapter11/default_cc_clients.csv')
```

Inspect this DataFrame with the `info()` command:

```
In [ ]:
    df.info()
Out[ ]:
    <class 'pandas.core.frame.DataFrame'>
    RangeIndex: 30000 entries, 0 to 29999
    Data columns (total 24 columns):
    LIMIT_BAL                    30000 non-null int64
    SEX                          30000 non-null int64
    EDUCATION                    30000 non-null int64
    MARRIAGE                     30000 non-null int64
    AGE                          30000 non-null int64
    PAY_0                        30000 non-null int64
    ...
    PAY_AMT6                     30000 non-null int64
    default payment next month   30000 non-null int64
    dtypes: int64(24)
    memory usage: 5.5 MB
```

The output is truncated, but the summary shows that we have 30,000 rows of credit default data available with 23 features. The target variable is the last column named `default payment next month`. A value of 1 indicates a default has occurred, and 0 otherwise.

Should you get a chance to open the CSV file, you will notice that all values in the dataset are in numeric format, and values such as gender, education, and marital status are already converted to the integer equivalent, saving the need for additional data preprocessing steps. Should you have datasets containing string or Boolean values, remember to perform label encoding and convert them to dummy or indicator values.

Splitting and scaling the data

Before feeding the dataset into our model, we have to prepare it in a proper format. The following steps guide you through the process:

1. Split the dataset into independent and target variables:

```
In [ ]:
    feature_columns= df.columns[:-1]
    features = df.loc[:, feature_columns]
    target = df.loc[:, 'default payment next month']
```

Our target values in the last column of the dataset are assigned to the target variable, while remaining values are feature values and are assigned to the features variable.

2. Split the dataset into training data and testing data:

```
In [ ]:
    from sklearn.model_selection import train_test_split

    train_features, test_features, train_target, test_target = \
        train_test_split(features, target, test_size=0.20,
random_state=0)
```

The train_test_split() command of sklearn helps to split arrays or matrices into random train and test subsets. Every non-keyword argument supplied provides a pair of train-test splits of input. Here, we will obtain two such pairs for input and output data. The test_size parameter indicates we will be including 20 percent of the input in the test split. The random_state parameter sets the random number generator to zero.

3. Convert the split data into NumPy array objects:

```
In [ ]:
    import numpy as np

    train_x, train_y = np.array(train_features),
np.array(train_target)
    test_x, test_y = np.array(test_features), np.array(test_target)
```

4. Finally, standardize the dataset by scaling the features with `MinMaxScaler()` of the `sklearn` module:

```
In [ ]:
    from sklearn.preprocessing import MinMaxScaler

    scaler = MinMaxScaler()
    train_scaled_x = scaler.fit_transform(train_x)
    test_scaled_x = scaler.transform(test_x)
```

As in the previous section, the `fit_transform()` and `transform()` commands are applied. However, this time the default scaling range is 0 to 1. With our dataset prepared, we can start to design a neural network using Keras.

Designing a deep neural network with five hidden layers using Keras

Keras uses the concept of layers when working with models. There are two ways to do so. The simplest way is by using a sequential model for a linear stack of layers. The other is the functional API for building complex models such as multi-output models, directed acyclic graphs, or models with shared layers. This means that the tensor output from a layer can be used to define a model, or a model itself can become a layer:

1. Let's use the Keras library and create a `Sequential` model:

```
In [ ]:
    from keras.models import Sequential
    from keras.layers import Dense
    from keras.layers import Dropout
    from keras.layers.normalization import BatchNormalization

    num_features = train_scaled_x.shape[1]

    model = Sequential()
    model.add(Dense(80, input_dim=num_features, activation='relu'))
    model.add(Dropout(0.2))
    model.add(Dense(80, activation='relu'))
    model.add(Dropout(0.2))
    model.add(Dense(40, activation='relu'))
    model.add(BatchNormalization())
    model.add(Dense(1, activation='sigmoid'))
```

The add() method simply adds layers to our model. The first and last layers are input and output layers respectively. Each Dense() command creates a regular layer of densely connected neurons. In between them, a dropout layer is used to randomly set input units to zero, helping to prevent overfitting. Here, we specified the dropout rate as 20%, though 20% to 50% is usually used.

The first Dense() command parameter with a value of 80 refers to the dimensionality of the output space. The optional input_dim parameter refers to the number of features for the input layer only. The ReLU activation function is specified for all except the output layer. Right before the output layer, a batch normalization layer transforms the activation mean to zero and standard deviation close to one. Together with a sigmoid activation function at the final output layer, output values can be rounded off to the nearest 0 or 1, satisfying our binary classification solution.

2. The summary() command prints a summary of the model:

```
In [ ]:
    model.summary()
Out[ ]:
```

Layer (type)	Output Shape	Param #
dense_17 (Dense)	(None, 80)	1920
dropout_9 (Dropout)	(None, 80)	0
dense_18 (Dense)	(None, 80)	6480
dropout_10 (Dropout)	(None, 80)	0
dense_19 (Dense)	(None, 40)	3240
batch_normalization_5 (Batch	(None, 40)	160
dense_20 (Dense)	(None, 1)	41

```
Total params: 11,841
Trainable params: 11,761
Non-trainable params: 80
```

We can see the output shape and weights in each layer. The number of parameters for a dense layer is calculated as the total number of weights matrix plus the number of elements in the bias matrix. For example, the first hidden layer, dense_17, will have 23×80+80=1920 parameters.

 The list of activations available in Keras can be found at https://keras.io/activations/.

3. Configure this model for training with the compile() command:

```
In [ ]:
    import tensorflow as tf

    model.compile(optimizer=tf.train.AdamOptimizer(),
                  loss='binary_crossentropy',
                  metrics=['accuracy'])
```

The optimizer parameter specifies the optimizer for training the model. Keras provides some optimizers, but we can choose a custom optimizer instance instead, using the Adam optimizer in TensorFlow from the previous section. The binary cross-entropy calculation is chosen as the loss function as it is suitable for our binary classification problem. The metrics parameter specifies a list of metrics to be produced during training and testing. Here, the accuracy will be produced for retrieval after fitting the model.

 A list of optimizers available in Keras can be found at https://keras.io/optimizers/. A list of loss functions available in Keras can be found at https://keras.io/losses/.

4. Now is the time to train our model using the fit() command with 100 epochs:

```
In [ ]:
    from keras.callbacks import History

    callback_history = History()

    model.fit(
        train_scaled_x, train_y,
        validation_split=0.2,
        epochs=100,
        callbacks=[callback_history]
    )
```

```
Out [ ]:
    Train on 19200 samples, validate on 4800 samples
    Epoch 1/100
    19200/19200 [==============================] - 2s 106us/step -
loss: 0.4209 - acc: 0.8242 - val_loss: 0.4456 - val_acc: 0.8125
...
```

The preceding output is truncated as the model produces detailed training updates for every epoch. A History() object is created and fed into the model's callback for recording events during training. The fit() command allows the number of epochs and batch size to be specified. The validation_split parameter is set such that 20% of the training data will be set aside as validation data, evaluating the loss and model metrics at the end of each epoch.

 Instead of training the data all at once, you can also train your data in batches. Call the fit() command with an epochs and batch_size parameter, like this: model.fit(x_train, y_train, epochs=5, batch_size=32). You can also train batches manually using the train_on_batch() command, like this: model.train_on_batch(x_batch, y_batch).

Measuring the performance of our model

Using our test data, we can compute the loss and accuracy of our model:

```
In [ ]:
    test_loss, test_acc = model.evaluate(test_scaled_x, test_y)
    print('Test loss:', test_loss)
    print('Test accuracy:', test_acc)
Out [ ]:
    6000/6000 [==============================] - 0s 33us/step
    Test loss: 0.432878403028
    Test accuracy: 0.824166666667
```

Our model has 82% prediction accuracy.

Running risk metrics

In Chapter 10, *Machine Learning for Finance*, we discussed the confusion matrix, accuracy score, precision score, recall score, and F1 score in measuring classification-based predictions. We can reuse those metrics on our model as well.

Since the model output is in the normalized decimal format between 0 and 1, we round it up to the nearest 0 or 1 integer to obtain the predicted binary classification labels:

```
In [ ]:
    predictions = model.predict(test_scaled_x)
    pred_values = predictions.round().ravel()
```

The ravel() command presents the result as a single list stored in the pred_values variable.

Compute and display the confusion matrix:

```
In [ ]:
    from sklearn.metrics import confusion_matrix

    matrix = confusion_matrix(test_y, pred_values)
In [ ]:
    %matplotlib inline
    import seaborn as sns
    import matplotlib.pyplot as plt

    flags = ['No', 'Yes']
    plt.subplots(figsize=(12,8))
    sns.heatmap(matrix.T, square=True, annot=True, fmt='g', cbar=True,
        cmap=plt.cm.Blues, xticklabels=flags, yticklabels=flags)
    plt.xlabel('Actual')
    plt.ylabel('Predicted')
    plt.title('Credit card payment default prediction');
```

This produces the following output:

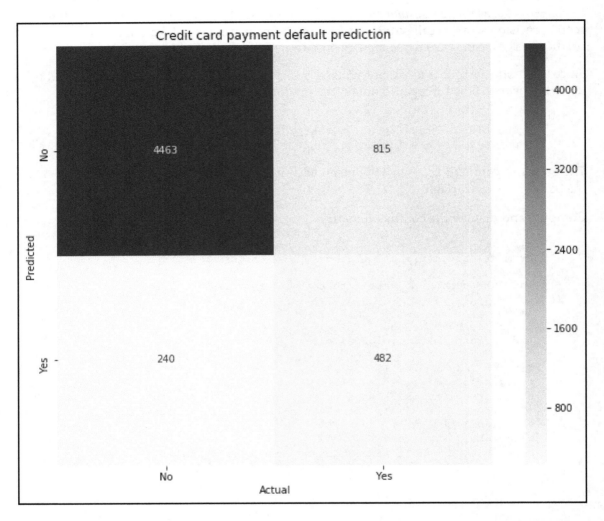

Print the accuracy, precision score, recall score, and F1 score using the `sklearn` module:

```
In [ ]:
    from sklearn.metrics import (
        accuracy_score, precision_score, recall_score, f1_score
    )
    actual, predicted = test_y, pred_values
    print('accuracy_score:', accuracy_score(actual, predicted))
    print('precision_score:', precision_score(actual, predicted))
```

```
      print('recall_score:', recall_score(actual, predicted))
      print('f1_score:', f1_score(actual, predicted))
Out[ ]:
      accuracy_score: 0.818666666667
      precision_score: 0.641025641026
      recall_score: 0.366229760987
      f1_score: 0.466143277723
```

The low recall score and the slightly below-average F1 score hint that our model is not sufficiently competitive. Perhaps we can visit historical metrics in the next section to find out more.

Displaying recorded events in Keras history

Let's review the `callback_history` variable, which is the `History` object populated during the `fit()` command. The `History.history` attribute is a dictionary containing four keys, storing the accuracy and loss values during training and validation. These are represented as a list of values saved after every epoch. Extract this information into separate variables:

```
In [ ]:
      train_acc = callback_history.history['acc']
      val_acc = callback_history.history['val_acc']
      train_loss = callback_history.history['loss']
      val_loss = callback_history.history['val_loss']
```

Plot the training and validation loss with the following codes:

```
In [ ]:
      %matplotlib inline
      import matplotlib.pyplot as plt

      epochs = range(1, len(train_acc)+1)

      plt.figure(figsize=(12,6))
      plt.plot(epochs, train_loss, label='Training')
      plt.plot(epochs, val_loss, '--', label='Validation')
      plt.title('Training and validation loss')
      plt.xlabel('epochs')
      plt.ylabel('loss')
      plt.legend();
```

This produces the following graph of losses:

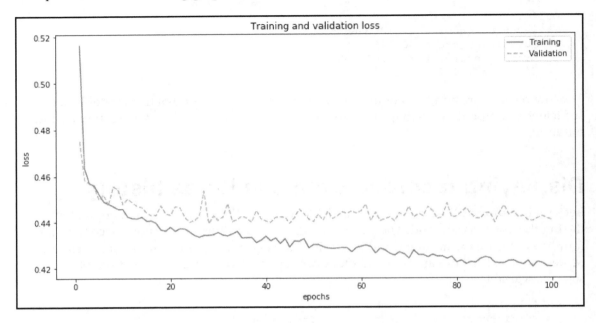

The solid line shows the path of the training loss decreasing as the number of epochs increases, meaning that our model is learning the training data better over time. The dashed line shows the validation loss increasing as the number of epochs increases, meaning that our model is not generalizing well enough on the validation set. These trends suggest that our model is prone to overfitting.

Plot the training and validation accuracy with the following code:

```
In [ ]:
    plt.clf()   # Clear the figure
    plt.plot(epochs, train_acc, '-', label='Training')
    plt.plot(epochs, val_acc, '--', label='Validation')
    plt.title('Training and validation accuracy')
    plt.xlabel('epochs')
    plt.ylabel('accuracy')
    plt.legend();
```

This produces the following graph:

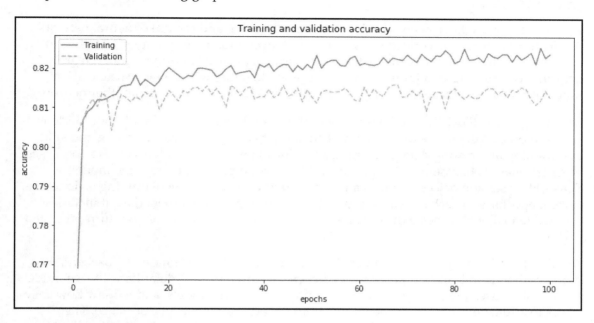

The solid line shows the path of the training accuracy increasing as the number of epochs increases, while the dashed line shows the validation accuracy decreasing. These two graphs strongly suggest that our model is overfitting the training data. Looks like more work needs to be done! To prevent overfitting, you can use more training data, reduce the capacity of the network, add weight regularization, and/or use a dropout layer. In reality, deep learning modeling requires understanding the underlying problem, finding a suitable neural network architecture, and investigating the effects of activation functions at each layer in order to produce good results.

Summary

In this chapter, we have been introduced to deep learning and the use of neural networks. An artificial neutral network consists of an input layer and an output layer, with one or more hidden layers in between. Each layer consists of artificial neurons, and each artificial neuron receives weighted inputs that are summed together with a bias. An activation function transforms these inputs into an output, and feeds it as input to another neuron.

Using the TensorFlow Python library, we built a deep learning model with four hidden layers to predict the prices of a security. The dataset is preprocessed by scaling and split into training and testing data. Designing an artificial neuron network involves two phases. The first phase is to assemble the graph, and the second phase is to train the model. A TensorFlow session object provides an execution environment, where training is done over several epochs, and each epoch uses mini-batch training. As the model output includes normalized values, we scale the data back to its original representation to return predicted prices.

Another popular deep learning library is Keras, utilizing TensorFlow as the backend. We built another deep learning model to predict credit card payment defaults with five hidden layers. Keras uses the concept of layers when working with models, and we saw how easy it was to add layers, configure the model, train it, and evaluate its performance. The History object of Keras records the loss and accuracy of training and validation data for successive epochs.

In reality, a good deep learning model requires effort and understanding the underlying problem in order to produce good results.

Other Books You May Enjoy

If you enjoyed this book, you may be interested in these other books by Packt:

Hands-On Python for Finance
Krish Naik

ISBN: 9781789346374

- Clean financial data with data preprocessing
- Visualize financial data using histograms, color plots, and graphs
- Perform time series analysis with pandas for forecasting
- Estimate covariance and the correlation between securities and stocks
- Optimize your portfolio to understand risks when there is a possibility of higher returns
- Calculate expected returns of a stock to measure the performance of a portfolio manager
- Create a prediction model using recurrent neural networks (RNN) with Keras and TensorFlow

Hands-On Machine Learning for Algorithmic Trading
Stefan Jansen

ISBN: 9781789346411

- Implement machine learning techniques to solve investment and trading problems
- Leverage market, fundamental, and alternative data to research alpha factors
- Design and fine-tune supervised, unsupervised, and reinforcement learning models
- Optimize portfolio risk and performance using pandas, NumPy, and scikit-learn
- Integrate machine learning models into a live trading strategy on Quantopian
- Evaluate strategies using reliable backtesting methodologies for time series
- Design and evaluate deep neural networks using Keras, PyTorch, and TensorFlow
- Work with reinforcement learning for trading strategies in the OpenAI Gym

Leave a review - let other readers know what you think

Please share your thoughts on this book with others by leaving a review on the site that you bought it from. If you purchased the book from Amazon, please leave us an honest review on this book's Amazon page. This is vital so that other potential readers can see and use your unbiased opinion to make purchasing decisions, we can understand what our customers think about our products, and our authors can see your feedback on the title that they have worked with Packt to create. It will only take a few minutes of your time, but is valuable to other potential customers, our authors, and Packt. Thank you!

Index

finite difference base class, writing 121
  implicit method 125, 126
  pricing American options 132, 133, 135, 136
fixed strategy parameters 318
fixed-income securities 144
forward rates 153, 154

G

gamma formula 107
Gauss-Seidel method 64, 65, 66
general nonlinear solvers 87, 88
Genetic Algorithm (GA) 326
good-till-canceled (GTC) order 268
gradient descent 362
gradient tree boosting regression model 352
Greeks 106
Greenwich Mean Time (GMT) 239
grid search
  model parameters, finding by 213

H

Hannan-Quinn information criterion 199
high learning rate 362
high-frequency trading (HFT) 264
histogram
  plotting 26
HyperText Transfer Protocol (HTTP) 266

I

implied volatility model 70
implied volatility modeling
  about 137
  of AAPL American put option 137, 138
in-the-money options (ITM) 71
incremental search 75, 76, 77
independent and identically-distributed (i.i.d) 72
integer programming
  about 51
  minimization example 51, 52
  with binary conditions 53, 55
Intrinio
  reference 372
IPython Notebook 12

J

Jacobi method 62, 63, 64
Jupyter Notebook
  running 12

K

k-means clustering algorithm 325
k-nearest neighbors (k-NN) classifier 363
k-nearest neighbors (KNN) 325
Keras
  about 384
  installing 384
  used, for designing deep neural network with five
    hidden layers 387, 388, 389
kernel density estimate (KDE) 216
kernel PCA
  applying 192
  performing 193
Kwiatkowski–Phillips–Schmidt–Shin (KPSS) tests
  211

L

lattices
  in option pricing 113
Least Absolute Shrinkage and Selection Operator
    (LASSO) regression 345
least-squares regression
  performing, on CAPM model 43
Leisen-Reimer (LR) tree
  using 103, 104
limit order 268
linear discriminant analysis (LDA) 363
linear equations
  solving, matrices used 55, 56
linear integer programming problem 51
linear models
  reference 345
linear optimization 47
linear programming
  maximization example 48, 50
linear programs
  outcomes 50
LinearRegression, of scikit-learn
  reference 338

Wijngaarden-Dekker-Brent method 85

CPSIA information can be obtained
at www.ICGtesting.com
Printed in the USA
LVHW062144060820
662518LV00021B/2390